RATINGS ANALYSIS
Theory and Practice

COMMUNICATION TEXTBOOK SERIES
Jennings Bryant—Editor

Broadcasting
James Fletcher—Advisor

BEVILLE • Audience Ratings: Radio, Television, Cable (Revised Student Edition)

ADAMS • Social Survey Methods for Mass Media Research

NMUNGWUN • Video Recording Technology: Its Impact on Media and Home Entertainment

MacFARLAND • Contemporary Radio Programming Strategies

WEBSTER/LICHTY • Ratings Analysis: Theory and Practice

RATINGS ANALYSIS
Theory and Practice

James G. Webster
Lawrence W. Lichty
Northwestern University

LAWRENCE ERLBAUM ASSOCIATES, PUBLISHERS
1991   Hillsdale, New Jersey   Hove and London

Copyright © 1991 by Lawrence Erlbaum Associates, Inc.
   All rights reserved. No part of the book may be reproduced in
   any form, by photostat, microform, retrieval system, or any other
   means, without the prior written permission of the publisher.

Lawrence Erlbaum Associates, Inc., Publishers
365 Broadway
Hillsdale, New Jersey 07642

Library of Congress Cataloging-in-Publication Data
Webster, James G.
   Ratings analysis : theory and practice / James G. Webster,
Lawrence W. Lichty.
   p. cm. — (Communication textbook series)
   Includes bibliographical references and index.
   ISBN 0-8058-0239-8 (c) / ISBN 0-8058-0949-x (p)
   1. Television programs—Ratings—Methodology. 2. Television
audiences. 3. Radio programs—Ratings—Methodology. 4. Radio
audiences. I. Lichty, Lawrence Wilson. II. Title. III. Series.
HE8700.65.W42   1991
384.54′3—dc20   90-40151
                                                            CIP

Printed in the United States of America
10 9 8 7 6 5 4 3 2 1

CONTENTS

Preface — vii

PART I: APPLICATIONS

Chapter 1   Ratings Analysis in Advertising   3

Chapter 2   Ratings Analysis in Programming   25

Chapter 3   Ratings Analysis in Social Science   42

PART II: RATINGS DATA

Chapter 4   The Ratings Research Business   67

Chapter 5   Ratings Research Methods   84

Chapter 6   Ratings Research Products   119

PART III: ANALYTICAL TECHNIQUES

Chapter 7   A Framework for Analysis   145

Chapter 8   Analysis of Gross Measures   185

Chapter 9	Analysis of Cumulative Measures	215
	Appendix A: National Ratings Research Companies	238
	Appendix B: Glossary	244
	Appendix C: ADI Market Rankings	261
	References	264
	Author Index	279
	Subject Index	281

PREFACE

This book was written with two groups of people in mind. First, it is intended for anyone who needs more than a superficial understanding of audience ratings data. This would certainly include many people who work in advertising, the electronic media, and related industries. For them, audience ratings are a fact of life. Whether they have been specifically trained to deal with ratings or not, their jobs typically require them to make use of "the numbers" when they buy and sell audiences, or make programming decisions. Also included in this category are students considering careers in the media—be they majors in broadcasting, marketing, communications, journalism, or other arts and sciences. For all these potential readers, the book might be entitled *Everything You Ever Wanted to Know about Audience Ratings*. We recognize that, for them, ratings analysis might be thought of as a necessary evil, so we have tried to make the book as plain spoken as our subject matter allows.

The second group of people for whom we have written this book includes those who are not compelled to use ratings data, but who nevertheless should. In this group we would include social scientists interested in mass communication, as well as those responsible for developing media policy in the United States. Although not wanting to sound like boosters for the ratings services, we believe that the data these companies collect offer rich possibilities for analysis that go well beyond the purposes for which they were collected. Indeed, ratings data can be thought of as offering up "texts" that clever analysts can "read" for insights into the social and economic impact of electronic media.

With these audiences in mind, we organized the book into three major sections. The first section illustrates the many applications of the ratings data. We spend a good deal of time discussing how the ratings are used in both advertising and programming (chapters 1 and 2). Such applications are, after all, the reason why we have ratings in the first place. Chapter 3 considers the use of ratings data in the social sciences, ranging from relatively pragmatic exercises in financial and economic analysis to questions of media effects and public policy.

The second section focuses on ratings data and the means by which they are collected. We chose to begin in chapter 4 with a brief history of the ratings business, not as an end in itself, but because we believe that understanding the forces that created audience ratings can contribute to a more perceptive view of the industry's present and future. This chapter also serves to introduce the major methods used to collect audience data. Chapter 5 describes the current methods of ratings research. Here we spend a good deal of time acquainting readers in the basics of sampling and survey research. Again, not as an end in itself, but because it is the key to understanding the strengths and limitations of the data offered for sale. Chapter 6 offers the reader a sampler of the products that the ratings services actually sell.

The final section of the book concentrates on the actual analysis of ratings data. Chapter 7 develops a theoretical model for analysis. Here, we have drawn together an eclectic mix of work from social psychology, marketing, and economics. The model is intended to embody what we know about audience behavior, and to offer a framework for further research—both applied and theoretical. In this we make a distinction between "gross" and "cumulative" measures of the audience. These categories organize the last two chapters of the book, each offering many examples of their respective sorts of analysis.

ACKNOWLEDGMENTS

We are indebted to many people for helping to make this book a reality. Some agreed to be interviewed, others reviewed drafts of the manuscript, still others helped out by providing the materials needed to illustrate various facets of ratings analysis. We value each contribution, and would like to thank George Bailey, Tom Bolger, Cynthia Brumfield, Jennings Bryant, Shelly Cagner, Ray Carroll, Ed Cohen, Alen Cooper, Rick Ducey, Andrew Ehrenberg, Jim Fletcher, Larry Frerk, David Giovannoni, Marcia Glauberman, Jack Hanrahan, David LeRoy, David MacFarland, Gale Metzger, Bill Miller, Peter Miller, Miriam Murphy, Bob Pepper, Pat Phalen, Joe Philport, Clark Pollack, Brad Rawlins, Rip Ridgeway, Jayne Zenaty Spittler, and Steve Wildman. Much of what is good about this book is a credit to them. The bad, we managed to introduce in spite of their help.

We would also like to acknowledge two very important, if different, sources of support in the preceding years. First, we are indebted to the Annenberg Washington Program in Communication Policy Studies and the Ameritech Foundation for their generous support of research projects that contributed directly to the writing of this book. Finally, we would like to thank our families, and most especially our wives, Debra and Sandra, for putting up with a couple of self-absorbed academics.

James G. Webster
Lawrence W. Lichty

Lawrence W. Lichty (left) James G. Webster (right)

PART I

APPLICATIONS

RATINGS ANALYSIS IN ADVERTISING 1

Audience ratings are a fact of life for virtually everyone connected with the electronic media. They are the tools used by advertisers and broadcasters to buy and sell audiences. They are the report cards that lead programmers to cancel some shows and to clone others. Ratings are also road maps to our patterns of media consumption, and as such, might be of interest to anyone from a Wall Street banker to a social scientist. They are the object of considerable fear and loathing, and they are certainly the subject of much confusion. We hope this book can end some of that confusion, and can lead to an improved understanding of both the ratings and the ways in which they can be analyzed.

It should be noted from the outset, that we use the term *ratings* as shorthand for a body of data on people's exposure to electronic media. Strictly speaking, ratings are one of many audience summaries that can be derived from that data. Specifically, ratings are estimated percentages of the population that see a program or listen to a station. Although these estimates, alone, occupy the attention of many media professionals, it is the full range of analytical possibilities offered by the larger database that is the subject of this book.

The best way to appreciate how ratings data can be analyzed is to become better acquainted with those who use the data, and the kinds of questions they are trying to answer. As is seen here, ratings data have significant uses in both programming and social science, but the most persuasive and important application of ratings is in advertising. So that is where we begin.

Broadcasters sell audiences. Despite some appearances to the contrary, that is the single most important activity of the business. Virtually all other actions are undertaken in support of that function. Whether this is good or bad can be, and frequently is, debated. For now, it is sufficient to note that this is an essential characteristic of commercial mass media. Not only do traditional broadcasters sell audiences, but newer forms of electronic media have also gotten into the act. Most cable networks, for example, will offer their audiences for sale.

The people who buy these audiences, of course, are advertisers. They are interested in capturing the attention of the viewer or listener in order to get across some message. It might be as simple as introducing people to a new name or reminding them of an old one. It might involve trying to change their attitudes toward a person or product. Often it represents an attempt to influence their behavior in some way. Whatever the advertiser's purpose, the process requires that they gain access to an audience—if only for a moment. In order to do that, they are willing to pay the media.

The difficulty with electronic media is that its audience has a unique, intangible quality. Unlike the print media, which can document readers with concrete figures on the number of issues they sell, broadcasters have to rely on estimates of who is out there listening. How those estimates are made is discussed in the next section of the book. Suffice it to say, that it was the desire of advertisers to buy audiences, and the eagerness of broadcasters to sell, that brought the ratings services into being. In that sense, this first category of ratings users is unique. No other group has a bigger stake in the ratings business. No other users wield more influence in shaping the form of the ratings. Indeed, were it not for the advertiser support of electronic media, ratings as we know them would not exist.

The buying and selling of audiences goes on at many different levels. There is a large national marketplace dominated by a few broadcast networks and major corporate advertisers. There are a great many local markets where individual stations sell to area merchants. And there are national spot and barter markets that provide access to audiences in various geographic regions. This trade in audiences is commonly organized by medium into radio, television, and cable. Table 1.1 gives some idea of how these markets have grown by summarizing the total revenues that have flowed to each medium as a result of time sales. All in all, its a multi-billion dollar business.

Each marketplace has developed its own institutions and practices. These characteristics can affect how ratings data are handled and the analytical techniques that are, or are not, employed. What follows is a description of each of the major marketplaces.

RATINGS ANALYSIS IN ADVERTISING

TABLE 1.1
Advertising Revenues of Electronic Media[a]

	Radio[b]			Television[c]				Cable[d]		Total
Year	Network	Spot	Local	Network	Spot	Local	Synd.	Network	Spot/Local	
1950	$132	$119	$203	$85	$31	$55	$—	$—	$—	$625
1960	45	208	402	820	527	280	—	—	—	2,282
1970	49	355	853	1,658	1,234	704	—	—	—	4,853
1980	158	746	2,643	5,130	3,269	2,967	50	50	8	15,021
1981	196	854	3,007	5,575	3,746	3,368	75	105	17	16,943
1982	218	909	3,365	6,120	4,364	3,765	150	195	32	19,208
1983	254	1,023	3,739	7,017	4,827	4,345	300	331	50	21,885
1984	288	1,184	4,412	8,526	5,488	5,084	430	486	86	25,924
1985	329	1,320	4,915	8,285	6,004	5,714	540	612	139	27,857
1986	380	1,333	5,313	8,570	6,570	6,514	610	740	192	30,222
1987	371	1,315	5,605	8,500	6,846	6,833	762	883	264	31,379
1988	382	1,402	6,109	9,172	7,147	7,270	901	1,061	351	33,795
1989	427	1,530	6,463	9,260	7,400	7,775	1,215	1,401	562	36,033

[a] Revenue in millions
[b] Radio Advertising Bureau
[c] Television Bureau of Advertising
[d] Cabletelevision Advertising Bureau

NETWORK SALES

The largest audiences, and the biggest single sums of money, are exchanged at the network level. Although radio and cable television offer network services, the major *television networks* have, far-and-away, the largest audiences. For advertisers who need to reach vast national markets, network television has much to offer.

As a practical matter, the network television marketplace is divided into a number of smaller markets. These are referred to as *dayparts*. All broadcasters divide their schedules into time periods they call dayparts. The precise name and definition of each daypart varies from medium to medium, and time zone to time zone. For the networks, a daypart is a portion of the their broadcast schedule defined both by time of day and by program content. These designations are useful because each one is associated with different audience characteristics. For that reason, different dayparts appeal to different advertisers and generate different amounts of money for the networks.

Prime time is the most important of the network dayparts. Unlike the official definition of prime time used by federal regulators, network prime time includes all regularly scheduled programs from 8 p.m. to 11 p.m. (ET), Monday through Saturday, and 7 p.m. to 11 p.m. on Sunday. During this daypart the networks have their largest audiences and, accordingly, generate their largest revenues. This daypart has special appeal to advertisers who are trying to reach a wide variety of people across the entire nation. It is also the best time to reach people who work during the day. Access to this mass market, however, does not come cheaply. In 1991, 30-second commercials in this daypart generally cost between $100,000 and $150,000. The most popular prime-time programs, of course, are the most expensive.

Daytime is the second most lucrative daypart. For the networks, it extends from 7 a.m. to 4:30 p.m., Monday through Friday. The daytime audience is much smaller, and with the exception of the early news programs, disproportionately female. As a result, it appeals most to advertisers who are trying to reach women, particularly women who do not work outside the home. Companies selling household products like soap and food stuffs frequently buy in this time period, but for nowhere near the cost of prime time. Through most of the 1980's, the average 30-second spot cost a little more than $10,000.

Sports is a daypart defined strictly by program content. The most important sports programming for the networks is coverage of major league games like those of the NFL or NBA. As might be expected, these events attract audiences that are disproportionately male. That fact is suggested by the list of advertisers who buy most heavily in this daypart. They include breweries, car and truck manufacturers, and companies that sell automotive products. The cost of advertising in sports programming varies widely—mostly as a function of audience size—all the way up to the Super Bowl that can cost $700,000 per spot.

The *news* daypart is another network market defined more by program content than by simple time periods. It includes the network's evening news programs, weekend news programming, and news specials and documentaries. It does not include everything that the network news divisions produce, however. Excluded from this daypart are the morning news programs (considered daytime), and regularly scheduled primetime programs like "60 Minutes." The news daypart tends to attract an older audience. It is, therefore, especially appealing to companies that sell products like headache remedies and healthful foods. In 1991, a 30-second spot in the evening news could cost roughly $50,000—less when purchased in volume.

Late night runs from 11:30 p.m. (ET) through the rest of the evening and early morning, Monday through Friday. Its best known program is "The Tonight Show," which has dominated the time period for decades. Not surprisingly, the audience during this daypart is small and almost entirely adult in composition. Depending on the program, a 30-second spot costs from $10,000 to $30,000.

One of the most important markets from the point of view of public interest groups and government regulators is the *children's* daypart. Traditionally, this has included the Saturday and Sunday morning children's programs. A time period that critics once dubbed the "children's ghetto." It may also include weekday programming aimed at children. Although children watch a great deal of television at other times, from an advertiser's viewpoint, this daypart is the most efficient way to gain access to the child audience. The biggest buyers of time in this daypart are cereal and candy makers, and toy manufacturers. The cost of a 30-second spot can vary widely, as demand for advertising time is seasonal. Leading into Christmas, a spot might cost three times as much as it would in the months that follow.

The buying and selling of network time occurs in different stages throughout the year. These different rounds in the buying process are called the *upfront market,* the *scatter market,* and the *opportunistic market.*

The upfront market is the first round of buying. Each spring and summer, major advertisers tell the networks what kind of audiences they wish to buy in the forthcoming television season. The network salespeople respond with proposals of what audiences they will sell and at what prices. Obviously, the networks want to get as much money for their audiences as possible, whereas the advertisers want to get as many viewers for their dollars as they can. All this is complicated by the fact that no one can know exactly what audiences will be attracted to the shows in the fall line-up, especially the new shows.

The upfront market is the occasion for much high stakes gamesmanship, which David Poltrack (1983) had described in detail. When all is said and done, the major network advertisers have made commitments to buy large blocks of network time throughout the coming year. More than half of each network's "inventory" of available commercial minutes is likely to be sold in the upfront market.

Although this method of buying may tie up an advertiser's budgets for months to come, it affords them access to the pick of network's inventory. Because these companies are making long-term commitments to the network, they are also likely to get time at more favorable rates than will be available later in the year. In fact, to minimize the advertiser's risk, network's will typically guarantee that the total audience numbers they have sold will be delivered, even if that means running additional commercials for free.

The scatter market operates within a shorter time frame. Each television season is divided into quarters. In advance of each quarter, advertisers may wish to buy time for specific purposes. It could be, for example, that some limited campaign, not envisioned during the upfront buying, will require the purchase of additional network time. Because advertisers usually come to the scatter market with less flexibility, and because the networks will have already sold much of their inventory, this market often finds the buyer at a disadvantage. That negotiating disadvantage usually means higher costs to the advertiser.

The opportunistic market occurs as the television season progresses. Even though most of the networks' inventory is sold during the upfront and scatter markets, some is not. Further, deals that were agreed to in those markets may fall through. For example, a new series may falter and have to be changed, relieving advertisers of their commitment. Similarly, the network may pre-empt regularly scheduled programs. All these things leave holes in the network line-ups and create opportunities for savvy buyers and sellers to exploit. Sometimes events operate to the advantage of the network, sometimes to the advertiser.

Despite their clear domination of national television audiences, the broadcast networks are not the only way to reach across the country with a televised message. Since the early days of television, an alternative delivery system has been developing. Cable television uses a wire to distribute signals, instead of broadcasting them through the spectrum. It originally functioned to supplement the broadcast delivery system, by bringing signals to areas that had poor over-the-air reception. As such, the early systems were little more than glorified antennas. In fact, cable was referred to by the acronym "CATV"—short for community antenna television.

After years of struggling with government regulators and much financial uncertainty, cable television has emerged as an advertising medium in its own right. Because systems use a high capacity wire, called a *coaxial cable,* they typically have an abundance of channel space. That, in combination with the growth of communication satellites—which can send TV signals to many small and widely dispersed cable systems—has opened a door for new "network" services. The limits of the electromagnetic spectrum no longer constrain the number of television signals that can compete for the viewers' attention.

Since the late 1970s a number of entrepreneurs have exploited these technological changes to create program services commoly referred to as cable television networks. Table 1.2 lists national cable television networks, the kind of programming they specialize in, and the number of homes they reach with their signals. All of these services depend, at least in part, on advertising revenues to sustain their operations. Indeed, many are programmed in a way that attracts a particular kind of viewer—the kind that interests an advertiser. MTV, for example, is designed for teens and young adults, Nickelodeon for children, and Univision for Hispanics.

There are roughly 93 million television households in the United States and, as noted, even the largest cable networks have nowhere near that

TABLE 1.2
Advertiser-Supported Cable Networks[a]

Network	Potential Audience (TVHH in millions)	Content
ESPN	54.0	Sports-oriented; football (college and NFL), hockey, tennis, soccer, and others
CNN	53.2	24-hour news with in-depth reporting
USA Network	51.9	Wide variety of entertainment, children's programming, sports, and a variety of news
Nickelodeon	50.0	Children's programming
MTV	49.9	24-hour music channel with videos, concerts, news, and interviews
TNN (The Nashville Network)	49.0	country music, live entertainment, sports, talk shows, and news
Family Channel	48.4	Family entertainment, children's programming, sports, and a variety of news
Discovery Channel	46.8	Documentaries, nature, science programming
Lifetime	46.0	Women's programming
Arts & Entertainment Network	41.0	Performing arts, drama, documentaries, and music
Weather Channel	41.0	Weather on the international, national, and local levels
CNN Headline News	40.4	Repeated, constantly updated half-hour news summaries
VH-1	35.9	24-hour music video channel geared to 24- to 49-year-old market
TNT (Turner Network Television)	35.0	24-hour entertainment channel with movies and original programming
FNN (Financial News Network)	32.0	Stock market information and business news
BET (Black Entertainment Network)	26.0	Talk shows, family programming, music, and religious programming
Prime Network	23.0	Sports events distributed through regional networks
TLC (The Learning Channel)	18.0	Educational programming
Movietime Channel	18.0	24-hour promotional channel for local movie theaters
Prevue Guide	15.2	Cable programming guide
Travel Channel	15.0	24-hour information with feature stories
CNBC (Consumer News and Business Channel)	14.0	Consumer issues, business, and market news
Nostalgia Channel	8.0	News, early movies, TV series geared to viewers 45- and older
Comedy Channel	6.0	24-hour comedy programming
Univision (formerly SIN)	5.2	Spanish-language news, movies, sports, and children's programming
Galavision/ECO	2.0	24-hour Spanish news service; movies, and sports (on Sundays only)

[a] Adapted from Cabletelevision Advertising Bureau (1990).

number of households in their potential audience. Despite rapid growth in the number of homes that subscribe to cable service, it is only in about half of all television households. Further, even optimistic estimates of penetration rates in the 1990s rarely exceed 70%. Although that includes a very large number of potential viewers, no cable network is likely to attract a national audience in excess of a broadcast network. Some viewers are simply beyond their reach.

These facts of life affect the way cable networks go about selling their audiences. One of three rationales is commonly used to appeal to advertisers. First, because the growth of cable has caused a steady decline in the amount of time people spend with broadcast television, cable networks often sell themselves as a way to get at those lost viewers. The sales pitch is that broadcast networks underdeliver the audience, and that buying time on cable networks corrects that problem in a cost-effective manner. Further, it is argued that cable households, where underdelivery is the biggest problem, include the most affluent and generally desirable target audiences. Second, because many cable services are designed to cater to specific subsets of the mass audience, advertising on the appropriate services is said to be a more efficient way to reach the kind of viewer an advertiser wants. Third, cable networks are often more willing to work with an advertiser to develop some special programming or promotional effort. This can sometimes enhance the impact of the advertising. A cable network, therefore, may not be able to sell an advertiser on the sheer size of its audience, but rather with the efficiencies and potential impact that the medium offers.

Although television networks, broadcast or cable, command much of our attention these days, it is worth remembering that the first networks were radio. Radio networks were permanently established by the late 1920s. They established many of the practices and traditions that are a part of network television today. In fact, radio networks have been an important social and cultural force in American life. Despite this rich history, radio networks are not what they used to be. Television has moved to center stage in our lives, and with it has come the lion's share of the advertising revenues. Nevertheless, radio networks are still very much with us, and offer advertisers another way of reaching a national audience.

There are about 20 radio networks, but many are controlled by three organizations. ABC Radio, for example, operates ABC Contemporary, ABC Direction, ABC Entertainment, ABC FM, ABC Information, ABC Rock, and ABC Talk. The same management can and does sell time on all of these networks. The other major network owners are Westwood One, which controls the Mutual and NBC networks, and United Stations, which bought the RKO radio networks. CBS, still a very important group owner of especially popular AM "all news" stations and FM hit and oldies music stations in the largest markets, operates two radio networks.

Much like cable television, radio networks tend to offer specialized programming designed to appeal to a certain kind of listener. In fact, because it is relatively inexpensive to produce radio programming, networks can fine tune the appeal of their services in a way that television finds difficult to match. That is one reason why ABC Radio can offer so many different types of radio formats. We have more to say about how radio programmers use ratings data to craft a format in the next chapter. One important consequence of this type of programming is that it allows advertisers to target their messages to specific audiences.

Unlike network television, however, advertisers who buy time on radio networks have particular concerns about whether their commercial is actually aired on all stations that are a part of the network. All network programming must be "cleared" by individual stations. In broadcast television, there are just a few major networks that are heavily depended on by affiliated stations. Because of this, all network commercials tend to be aired as planned. In radio, because there are more networks and affiliation is not so important, network commercials, sometimes, do not air as expected. A radio station may simply not clear the network programming, or it may run local ads over network spots. In fact, the need to monitor commercial clearance has become another chore that ratings services have been called on to do.

We should also note that although they there are not legally defined as *networks,* many radio programming syndicators provide specialized radio formats simultaneously via satellite to a large number of stations all over the country.

LOCAL SALES

Broadcast networks reach national markets by combining the audiences of the local stations with whom they affiliate. Similarly, cable networks aggregate the viewers of local cable systems. But individual stations and systems can and do sell audiences by themselves. Local audiences such as these offer advertisers a way to reach smaller, geographically concentrated markets. That appeals to a great many advertisers who only do business locally. In fact, national or regional advertisers often buy combinations of local markets, because it offers them greater flexibility than could be achieved through a network buy. All these can be considered local sales, and they constitute another marketplace for media audiences.

The physics of broadcasting are such that a station's signal has geographic limits. In light of this, the FCC decided to license radio and television stations to specific cities and towns across the country. Larger population centers have more stations. Naturally enough, people spend most of their time listening to stations in close proximity because they can receive the clearest signal and can hear some programs of local interest. The major ratings services use this geographically determined audience behavior to define the boundaries of a local media market area. Two schemes of market designation are in common use. Arbitron, one ratings firm, calls each market an "area of dominant influence" (ADI). A.C. Nielsen labels each area a "designated market area" (DMA). For the most part, ADIs and DMAs are the same.

There are over 200 of these markets in the United States. Appendix C lists the television markets designated by Arbitron. There are even more radio markets. In either case, market size varies considerably. New York, for instance has over 7 million TV households, whereas North Platt has fewer than 20,000. Indeed, buying time on a major station in New York, might deliver more viewers to an advertiser than a national cable network. Conversely, many small market radio stations might have audiences too small for a ratings company to economically measure. This point is best illustrated by the fact that regular radio ratings are available to only about 3,000 of more than 10,000 stations in the country.

These vast differences in audience size have a marked effect on the rates that local broadcasters can charge for a commercial spot. The price of a 30-second spot in prime time might be $400 in Des Moines and $4,000 in Detroit. Other factors can affect the cost of time, too. Is the market growing or has it fallen on hard times? Is the population relatively affluent or poor? How competitive are other local media like newspapers? Even things like a market's time zone can affect the rates of local electronic media.

Another thing that varies with market size is the sophistication of ratings users, and the sheer volume of audience information they must deal with. As we will discuss in chapter 6, many radio markets have their audiences measured just once a year. Major TV markets, on the other hand, have their audiences measured continuously. Because of this, and the greater number of advertising dollars available in major markets, the buyers and sellers of media in those markets tend to be more experienced with and adept at analyzing ratings information.

In most markets, the biggest buyers of local advertising include fast food restaurants, supermarkets, department stores, banks, and car dealers. Like network advertisers, these companies will often have an *advertising agency* represent their interests. The agency can perform a number of functions for its client, from developing a creative strategy, to actually writing copy and producing the ads. Most important in this context, the agency will project the audience for various programs, plan when the ads are to run, buy the time, and evaluate whether the desired audience was delivered. Smaller advertisers, or those in smaller markets, may deal directly with the local stations.

Because of the different types of people involved, the process of buying local time varies somewhat from market to market. It might be a "seat of the pants" judgment made by a merchant who believes that buying a certain number of ads on a local station generates extra business. Indeed, many small radio stations sell without using any ratings information at all. Increasingly, though, the process of buying and selling time is dependent on the use of ratings information.

Generally, the purchase of local time works like this. The advertiser or its agency will issue what's called a request for *avails*. In effect, the buyer is asking what spots are available for sale on the local stations. Avail requests will typically specify the kind of audience that buyers want, the daypart they wish to buy in, and will give some indication of the budget they have to spend. Station salespeople will respond by proposing a list of spots, called a *submission,* that will deliver some or all of the audience the buyer wants. At this point the buyer and seller enter a period of negotiation during which differences over the projected audience and its cost are ironed out. Assuming that the parties can reach an agreement, the buyer will place an order and the spots will be aired. After the campaign has run, the next available ratings information is used to determine whether the expected audience was actually delivered. This last stage in the process is called *post-buy analysis.*

National and regional advertisers also buy spots on local stations. For example, a snow tire manufacturer might want to advertise only in northern markets. Similarly, a maker of agricultural products might wish to buy time in markets with large farm populations. In fact, such *national spot* buys constitute the largest single source of revenues for many TV stations. The question is, how can so many local stations deal effectively with all these potential time buyers. It would be impractical for thousands of stations to have their own personnel trying to contact each and every national advertiser.

To solve this problem, an intermediary called a *station representative* serves as the link between local station inventories and national advertisers. Rep firms for both television and radio stations are located in major media markets like New York and Chicago. Television reps will have only one client per market, so as to avoid any conflict of interests. Radio reps may serve more than one station in a market, as long as their formats don't compete for the same audience. Rep firms vary in terms of the number of stations they work with, and the types of services they offer their clients. Some firms provide stations with research services, or advice on their programming. Most importantly, though, they determine what media buys national advertisers are planning, and try to secure some portion of that business for their stations. To do this requires constantly making the rounds of the major ad agencies and media buyers. As with other forms of media buying, the players must demonstrate good negotiating skills, and a facility for dealing with ratings data.

Cable systems, too, have the potential to offer local advertising, and some systems are starting to explore that potential. Usually, this means inserting a local ad in a cable network. One problem is that, just like cable networks, cable systems simply can't reach every member of the audience. Nevertheless, there is a different potential here that advertisers might be able to exploit.

In a way, cable system advertising is not just local, its "ultra-local." In most TV markets, for instance, there are several cable systems. Its possible that an advertiser, like a small merchant, would want to run a spot only in one or two communities within the larger market. Advertising on cable could do this. Similarly, since cable franchise areas, almost by definition, conform to governmental boundaries within the market, political advertising seems like a likely candidate for local cable advertising. Further, if several cable systems coordinate their efforts, rather precise and varied geographic coverage of the market is possible.

SYNDICATION

Stations are in constant need of programming. Even those that affiliate with a network have large blocks of time they must program themselves. As a result, broadcasters have had to acquire programming from different sources. One such source, which is particularly relevant to a discussion of advertising, has been barter syndication.

Barter syndication has fairly straight forward origins. Basically, advertisers found they could use a station's need for programming to get their message across to the audience. All they had to do was produce a program, place their ads in it, and offer it to stations free of charge. Stations found this attractive because they got new programs at no cost, and could even sell some spots in the show not used by the program's original sponsor. In the 1980's, with the advent of satellite program distribution, this simple idea gave rise to a rapidly growing new advertising marketplace.

Today, barter syndication works like this. A distributor that produces programming and/or owns the rights to existing programming, will go to local broadcasters and induce the stations to carry the show. Sometimes this is a traditional barter arrangement. Increasingly, though, its what's called "cash-plus-barter." Under this arrangement, the station actually pays a fee for the program, in addition to accepting ads placed by the distributor. Further, cash-plus-barter contracts may require the station to broadcast the program in a specific daypart. This sort of deal is typical of popular programs, like "Wheel of Fortune," that are especially desired by stations. In any case, the distributor of the program now has time to sell an advertiser.

The more stations that acquire a program, the larger is the potential audience. If one station in every market agreed to air the program, the distributor would, hypothetically, have the same reach as a major television network. As a practical matter, once a program is carried on enough stations to reach 70% of U.S. households, it is sold to advertisers much the same way that network time is sold.

A handful of firms dominate the sale of time in barter syndication. Each one represents a variety of distributors, and in fact, most have their own programming to sell. Table 1.3 lists these companies, some of the distributors they represent, and syndicated programming illustrative of each distributor.

These barter syndication firms, and other smaller companies, go to national advertisers and their ad agencies to sell time. Just like the networks, these companies sell in upfront, scatter, and opportunistic markets. Just like the networks, upfront sales include some guarantee of audience delivery. In fact, advertisers tend to look upon barter syndication as a supplement to their purchases of network time. Some of these sponsors

TABLE 1.3
Major Barter Sales Firms

	Distributor	Program
Camelot Entertainment Sales	King World	"Wheel of Fortune"
	King World	"Jeopardy!"
	King World	"Oprah Winfrey"
Premier Advertiser Sales	Paramount	"Star Trek: The Next Generation"
	Paramount	"Arsenio Hall Show"
Television Program	TPE	"Star Search"
Enterprises	Paramount/MCA	"Entertainment Tonight"
Tribune Entertainment	Tribune	"At the Movies"
	Paramount	"Geraldo"
Group W Productions Media	Viacom	"Cosby" (off network)
Sales	MGM	"New Twilight Zone"
Spectrum	Multimedia	"Donahue"
	Multimedia	"Sally Jessy Raphael"
TV Horizons	LBS	"Family Feud"
	Fox	"Current Affair"

are especially interested in barter because they can be assured of the "program environment" into which their messages will be placed. But the major attractiveness of barter is providing participation at a somewhat lower cost.

Despite these similarities, buying time in barter syndication is not quite comparable to network advertising. For one thing, there isn't the same real time delivery of the advertising message. It is common for a given program to air at different times in different markets. A syndicated program that is on once a week, might even be broadcast on different days. Advertisers also have lingering doubts about whether their commercials are actually clearing in all the markets showing the program. Finally, syndicators have less clout with stations than TV networks, therefore some broadcasters have been tempted to remove the commercials in syndicated programs in favor of their own local ads.

Problems or not, barter syndication and related ways to package advertising for national or regional audiences are almost certain to grow. Satellite communications have made the rapid, cost-efficient delivery of programming feasible. Program services, distributed to stations in this way, are, in effect, ad hoc networks. If it is to their advantage, stations will pick up these syndicated program feeds, perhaps even pre-empting more traditional networks. Assuming that there is an effective way to buy and evaluate the audiences, advertisers are likely to use these alternative routes for reaching the public. Such ever changing syndicated networks are also likely to pose some of the most interesting challenges for audience analysts.

RESEARCH QUESTIONS

Obviously, the buying and selling of audiences happens in a number of different places, and involves people with different motivations and levels of sophistication. There are, nonetheless, a handful of recurring research questions that transcend these differences. By distilling these from the foregoing discussion we can simplify what is going on, and see more clearly how ratings data are used in the context of sales and advertising. There are basically four questions these users of the ratings are trying to answer.

How Many People Are in the Audience? More than any single factor, it is the size of the media audience that determines its value to advertisers and, in turn, its value to broadcasters. There are a number of different ways to express audience size. In order to acquire a working vocabulary, we discuss the most common of these now, and leave more technical definitions until the last section of the book.

RATINGS ANALYSIS IN ADVERTISING

Ratings are the most frequently used descriptors of audience size. Indeed, this term is so widely recognized that we have titled the book with it. A rating is actually the percentage of the population that listen to a particular station or watch a particular program. The simplest version of a ratings calculation is presented in Fig. 1.1, as are other standard expressions of audience size.

Two characteristics of a rating should be noted. First, the population figure on which the rating is based is the total potential audience for the program or station. For local stations, that is usually the population in the market equipped with radios or television sets. For all intents and purposes, that's the entire population. It does not matter whether those sets are being used or not, the population estimate is the same for all ratings calculations. In this context, it means that the denominator of the ratings term does not vary from station to station, or program to program. To say a TV program had a rating of 20, then, means that 20% of the entire population in the market saw the show. Second, populations can be composed of different building blocks, or "units of analysis." In television, for example, it is common to talk about a population of television households. One "rating point," therefore, means one percent of the homes equipped with television in the given market area. In radio and, increas-

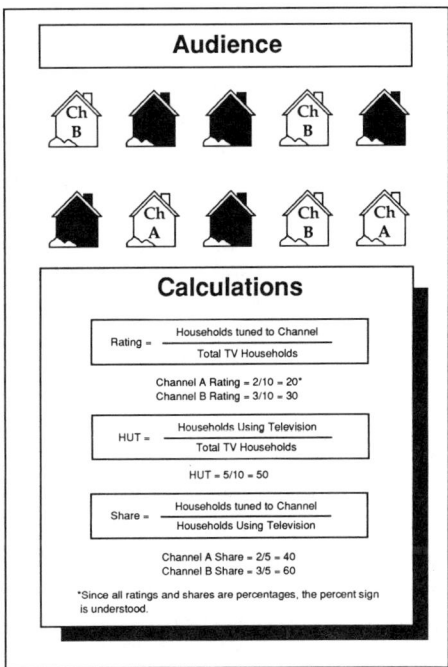

FIG. 1.1. Simple rating and share calculations.

ingly, in television, we also describe populations of people. Here, we might talk about a station's ratings among men or women of a certain age. As you could tell from our discussion of network dayparts, its quite possible for a program to have a relatively high rating among women and a small rating among men (e.g., daytime).

Another way to describe the size of the audience is to express is in absolute terms—the projected total size of the audience. Local radio audiences are usually counted in the hundreds of people. Television audiences are numbered in the thousands at the local level, and in the tens of thousands or millions at the network level. In some ways, absolute estimates of audience size are more interpretable. To know that a local station had a rating of 25, for example, gives you little idea how many human beings were actually in the audience unless you know the size of the market. Its quite possible, therefore, that a 25 rating in one market means a smaller audience than a 15 rating in a larger market.

Ratings and absolute numbers are just different expressions of audience size. They are based on the same data. Further, they are only estimates, not values we know to be perfectly accurate. In fact, much time and effort goes into collecting data on which reliable estimates can be based.

In addition to ratings, it is frequently useful to summarize the total number of people using the medium at any point in time. When households are the units of analysis, this summary is called "households using television," or *HUT* level for short. Figure 1.1 illustrates how this measure is calculated. As you will note, HUT levels are typically expressed as a percentage of the population. As with ratings, though, its possible to express them in absolute terms. If individuals are being counted, "persons using television," or *PUT,* is the appropriate term. In radio, the analogous terms are *PUR,* or *SIU* (i.e., sets in use), as each set is associated with an individual listener.

Not everyone uses television or listens to radio at the same time, so HUT levels will vary throughout the day. In fact, they change in a very predictable way, hour to hour, and week to week. Because of this, many audience analysts prefer to see a program's or station's audience expressed as a percentage of the HUT level, rather than the total population. Its as if they were saying, "Since I can't affect the size of the total audience in any given time period, just tell me how I did relative to the competition." The measure that expresses this is called an audience *share*. Figure 1.1 summarizes the share of audience calculation. It is quite possible, with this method of calculation, for a program to have a large share and a small rating. That would be true, for instance, for popular programs airing at times when very few people have their sets on. In fact, unless everyone is using the medium at the same time, a program's share will always be larger than its rating. It should also be apparent that shares, by themselves, give you no indication of the absolute size of an audience.

Shares are commonly used by buyers and sellers when they must predict the size of the audience. Actual ratings data are, by definition, historical. They describe what has already happened. Advertisers, by contrast, are trying to buy access to audiences that will be created at some time in the future. This means that they must make a projection about the likely audience. The standard way to do that, is to (a) determine what the HUT level will be when the spot is aired, and, (b) estimate the share of audience for the program containing the spot. This combination of factors allows one to project the size of the audience. More is said about these techniques in chapter 8. Advertisers will typically run a series of ads over a period of days or weeks. In some ways, then, the audience for a single airing of a commercial is less important than the total audience exposure to the ad campaign. To provide some assessment of the total audience exposed to the advertiser's message, the audience ratings associated with each individual commercial can be summed across all commercials in the campaign. This grand total is referred to as "gross ratings points," or GRPs. The term is used quite commonly in advertising, and almost nowhere else.

GRPs provide a crude measure of the total weight of a media campaign. In addition to summing audience ratings after the fact, the GRP concept is frequently used by advertisers to signal the total amount of audience they wish to buy from the media. For example, the avail request described earlier usually features a statement about the number of GRPs the buyer wants to accumulate in a particular campaign. GRPs, then, express the total size of the campaign in audience numbers rather than dollars.

One problem with GRPs is that they mask some very important features of audience behavior. For example, 100 GRPs could mean that 100% of the audience has seen a commercial just one time. However, it could also mean that one percent of the audience has seen the ad 100 times. Without further analysis, it's difficult to know what's happening underneath the veneer of GRPs.

How Often do the Same People Show Up in the Audience? To determine what audience behavior underlies GRPs, we need information on how each individual uses a medium over time. For example, we might want to know whether two programs with equal ratings were seen by the same people, or two entirely different groups of people. This is a question of "audience duplication." Fortunately, the same data that allow us to estimate gross measures of audience size like ratings, shares, and GRPs, also allow us to assess cumulative measures like audience duplication.

Advertisers are, understandably, interested in how many different people see their message, and how often they see it. These concerns are addressed by measures of reach and frequency, respectively. The *reach* of a commercial is defined by the total number of unduplicated individuals who are exposed to the ad. It is often expressed as a percentage of the total possible audience, just like a rating. In fact, there is a special kind of rating called a "cumulative rating," or *cume,* that measures the unduplicated audience for a station. In either case, the number represents the total number of different individuals who appear in the audience over some specified period of time.

Certain media are better at achieving large cumulative audiences than others. Prime time network television, for example produces considerable reach for a commercial message, since its audiences tend to be quite large. Further, many people only watch TV in prime time and, therefore, are reachable only in that daypart. As a result, advertisers are often willing to pay a premium for prime time spots. Cable networks, on the other hand, are limited by the penetration of cable systems, and so cannot hope to achieve penetration levels much in excess of 50%.

The second factor that comes into play is the *frequency* of exposure. The question here is, of those people who were reached, how many times did they actually see or hear the message. Measures of frequency provide the answer. Usually, information on the frequency of exposure is expressed as an average (e.g., "the average frequency was 2.8"). Of course, no one actually sees an ad 2.8 times, so it may be more useful for the advertiser to consider the full distribution on which the average is based. For example, if the advertiser believes that a person must see an ad three times before its effective, then he or she might want to know how many people saw the ad three or more times.

As was the case with reach, different media are better or worse at achieving a desired frequency. If, for instance, you wanted to market a product to Spanish-speaking audiences, buying time on an hispanic station might produce relatively low reach, but relatively high frequency. Similarly, radio can be an effective medium for achieving a high frequency of exposure, since the audiences for most stations tend to be loyal to station formats.

Reach and frequency bear a strict arithmetic relationship to GRP's. Specifically, reach multiplied by average frequency will equal gross rating points. If you know any two elements in this simple equation, you can derive the third. Unfortunately, advertisers usually know only GRP's, since only ratings are easily obtained from published reports. However, a number of agencies and audience researchers have developed mathematical models that will estimate reach and frequency given GRP's alone.

These and other techniques of modeling audience behavior are discussed in the last chapter.

Who Are the Audience Members? We have made references throughout this section of the book to the fact that different advertisers are interested in reaching different kinds of audiences. If the size of the audience is the most important determinant of its value, the composition of the audience is not far behind. In fact, advertisers are increasingly interested in presenting their messages to specific subsets of the mass audience. This strategy is referred to as market "segmentation." It plays a very important role in advertising and, in turn, has a major impact on the form that ratings data take.

Describing or segmenting an audience is accomplished by noting the characteristics or traits of the audience members. Researchers call these characteristics "variables." Almost any attribute can become a variable, as long as its reasonably well defined. In practice, viewer or listener attributes are usually grouped into one of four categories.

Demographic variables are the most commonly reported in ratings data. By convention, we include in this category such attributes as age, gender, income, education, marital status, and occupation. Of these, age and gender are far-and-away the most frequently reported audience characteristics. In fact, they form the basis of the standard reporting categories that are featured in ratings books. So, for example, advertisers and broadcasters will often buy and sell "women 18 to 49," "men 25 to 54," and so on.

Demographics have much to recommend them as segmentation variables. For one thing, everyone in the industry is used to working with them. When you talk about an audience of women or 18- to 34-year-olds, everybody knows exactly what you're talking about. On the other hand, there may be important differences between two men of the same age, differences that are potentially important to an advertiser. Therefore, additional methods of segmentation are used.

Geographic variables offer another common way to describe the audience. We have already encountered one of the most important, market areas. Just as people differ from one another with respect to their age and sex, so too, they differ in terms of where they live. Every TV viewer or radio listener in the country can be assigned to one particular market area. Obviously, such distinctions would be important to an advertiser whose goods or services had distinct regional appeal.

The other geographic variables that are commonly used in ratings research are, county and state of residence (including breakouts by county size), and region of the country. Recently, tracking a person's zip code has become a popular tool of geographic segmentation. With such finely drawn areas, it is often possible to make inferences about a person's income, lifestyle, and station in life. These zip code based techniques of segmentation are commonly referred to as "geo-demographics."

Behavioral variables draw distinctions among people on the basis of their behaviors. The most obvious kind of behavior to track is media use. We need to know who watched a particular program before we can estimate the size of its audience. With this kind of information, it is possible to describe an audience not only in terms of age and gender, but also in terms of what else they watched or listened to. Such audience breakouts, however, are only occasionally provided by the ratings service.

The other behavioral variables that weigh heavily in an advertisers mind are product purchase variables. When all is said and done, most advertisers want to reach the audience that is most likely to buy their product. What better way to describe an audience, then, than by purchase behaviors? For example, we could characterize an audience by the number of heavy beer drinkers it has, or the amount of laundry soap it buys. One ratings company has called such segmentation variables "buyer-graphics." As you might imagine, this is an approach to audience segmentation that has much appeal to advertisers.

The fourth category of variables that deserves brief mention here is *psychographics*. Definitions of this grouping vary, but basically it encompasses any attempt to draw distinctions among people on the basis of their psychological attributes. This would include things like people's values, attitudes, opinions, motivations, and preferences. Although such traits can, in principle, be very valuable in describing an audience, psychographic variables are often difficult to precisely define and measure.

How Much Does it Cost to Reach the Audience? All this time buying obviously costs money. Advertisers and the media, as well as middle men like ad agencies and station reps, all have an interest in what it costs to reach the audience. Those on the selling side of the business try to maximize their revenues, whereas buyers try to minimize their expenses.

Although it is true that broadcasters, and other forms of electronic media, sell audiences, it would be an oversimplification to suggest that audience factors alone determine the cost of a commercial spot. Certainly, audience size and composition are the principle determinants, but several factors have an impact. We have already pointed out that advertisers who buy network time early in the upfront market can get a better price. Similarly, advertisers who agree to buy large blocks of time can usually get some sort of quantity discount. Remember that these transactions happen in a marketplace environment. The relative strengths and weaknesses of each party, their negotiating skills, and, ultimately, the laws of supply and demand all affect the final cost of time.

These factors are represented in the rates that the media charge for commercial spots. It is common practice for an individual station to summarize these in a *rate card*. A rate card, which is usually presented in the form of a table or chart, will state the price of spots in different dayparts or programs.

Although the cost of a commercial spot is important to know, from the buyer's perspective, it is largely uninterpretible without associated audience information. The question the buyer must answer is, "What am I getting for the money?" This cannot be answered without comparing audience ratings to the rates that are being charged.

There are two common ways to make such comparisons. One technique is to calculate the cost per thousand (CPM) for a given spot (incidentally, the "M" in this expression is the Roman numeral for 1,000). To determine CPMs, you take the cost of commercial time and divide it by the size of the commercial audience (expressed in thousands). CPMs can be produced for households, women ages 18 to 49, or whatever kind of audience is most relevant to the advertiser. The calculation provides a yardstick to measure the efficiency of buying time on different stations or networks. The second method of comparison is to calculate a cost per point (CPP). Like CPM, this takes the cost of time and divides it by the number of ratings points delivered. Because ratings are not based on the same audience sizes across markets, the cost of a point will vary from market to market. Points in New York will be more expensive than points in Indianapolis. CPPs are useful, however, because they are easy to relate to GRPs. If, for example, an ad campaign is to produce 200 GRPs, and the CPP is $1,000, then the campaign will cost $200,000.

This sort of arithmetic reveals the economics that drive the industry. It is also a common form of ratings analysis among the buyers and sellers of time. But the media are complex organizations that can and do use audience information in a variety of ways. Similarly, those who want to study or regulate mass communication have found that the data gathered for the benefit of advertisers can offer many insights into the powers and potentials of the electronic media. These applications of the data are discussed in the chapters that follow. Then, we return to many of the concepts and specific applications that were discussed briefly earlier.

RELATED READINGS

Heighton, E. J., & Cunningham, D. R. (1984). *Advertising in the broadcasting media* (2nd ed.). Belmont, CA: Wadsworth.

Lavine, J. M., & Wackman, D. B. (1988). *Managing media organizations: Effective leadership of the media.* New York: Longman.

Lazer, W. (1987). *Handbook of demographics for marketing and advertising.* Lexington, MA: Lexington Books.
Marcus. N. (1986). *Broadcast and cable management.* Englewood Cliffs, NJ: Prentice-Hall.
Poltrack, D. F. (1983). *Television marketing: Network, local, cable.* New York: McGraw-Hill.
Sherman, B. L. (1987). *Telecommunications management: The broadcast & cable industries.* New York: McGraw-Hill.
Turow, J. (1984). *Media industries.* New York: Longman.
Warner, C. (1986). *Broadcast and cable selling.* Belmont, CA: Wadsworth.
White, B. C., & Satterthwaite, N. D. (1989). *But first these messages . . . The selling of broadcast advertising.* Boston: Allyn and Bacon.
Zeigler, S. K., & Howard, H. (1990). *Broadcast advertising* (3rd ed.). Ames, IA: Iowa.

RATINGS ANALYSIS IN PROGRAMMING   2

Following sales and advertising, the second most important application of ratings data is in programming. In order to sell commercial time, the electronic media must attract the audience that gives the time value. Broadly speaking, that is the job of a programmer. It is the programmer who sets the "bait" that lures the audience.

Programming involves a range of activities. A programmer must determine where the programming itself will come from. Sometimes that means playing an active role in developing new program concepts and commissioning the production of pilots. More often, it means securing the rights to syndicated programs, some of which have already been produced. In this capacity the programmer must be skillful in negotiating contracts and in divining what kinds of material will appeal to prospective audiences. Programmers are also responsible for deciding how and when that material will actually air on the station or network. Successful scheduling requires that a programmer know when different kinds of audiences are likely to be available, and how those audiences might decide among the options offered by competing media. Finally, a programmer must be adept at promoting the programming in the schedule. Sometimes that involves placing ads and "promos" to alert the audience to a particular program or personality. It can also involve packaging an entire schedule of programming in order to create a special station or network "image." In all such activities, ratings play an important role.

The way in which these programming functions are actually operationalized, and the priorities facing individual programmers, differ from one setting to the next. Occasionally, in small stations, the entire job of programming falls on the shoulders of one person.

In larger operations, however, programming will involve many people. In fact, because the media marketplace has become so competitive, the job of promoting programs and developing a certain image is increasingly turned over to specialized promotions departments.

The most significant difference in how programmers function, however, depends on the medium in which they work. Television, in the early 1950s, forced radio to adapt. No longer would individual radio programs dominate the medium. Instead, radio stations began to specialize in certain kinds of music or in continuous program formats. The job of a radio programmer became one of crafting an entire program service. Further, the vast supply of music from the record industry meant that stations could be less reliant on networks to define that service. Television, however, has built audiences by attracting them to individual programs. Although some cable networks now emulate radio by offering a steady diet of one type of programming (e.g., news, music, weather, financial and business information, or comedy), TV programmers, for the most part, must devote more attention to the acquisition, scheduling, and promotion of relatively discrete units of content. Much of this work has been done by broadcast networks. However, the growth of "independent" TV stations, as well as an evermore vigorous syndication market, has increased the amount of programming done by individual TV stations. To appreciate how ratings are used in programming, therefore, it is important to understand the programming practices of each medium.

RADIO PROGRAMMING

There are nearly 11,000 radio stations in the United States, each offering different—sometimes only slightly different—programming and each reaching different audiences. Most radio stations, from the smallest town to the largest markets, have a *format*. A format is an identifiable set of program presentations or style of programming. Some stations, particularly those in smaller markets with fewer competitors, may have a wider ranging format that tries to include a little something for everyone. Most stations, however, zero in on a fairly specific brand of talk or music.

Radio formats are important to the medium for two related reasons. First, radio tends to be a very competitive medium. In any given market, there will be far more radio stations than TV stations, daily newspapers, or almost any other local advertising medium you can think of. To avoid being lost in the shuffle, programmers have found that it helps to make their station seem unique or special. The strategy for doing that is sometimes called *positioning* the station. By differentiating their station from the others, programmers hope it will stand out in the minds of listeners, and induce them to tune in. Defining and promoting the right format is critical in positioning a station. Second, different formats are known to appeal to different kinds of listeners. Because most advertisers want to reach particular kinds of audiences, the ability to deliver a certain demographic is important in selling the station's time.

Radio formats run the gamut. Stations may call themselves "modern country," "continuous country," "all news," "news and talk," "new music," or even "kick ass rock and roll." Radio programmers, consultants, and analysts have fairly specific names for nearly 40 different formats. However, most of these are usually grouped in about a dozen categories. The most common labels, the number of stations and the share of the U.S. audience that listens to each is described in Table 2.1.

There are different ways to program a radio station. Some stations do all their own programming. They identify the specific songs they will play, and how often they will play them. They may also hire highly visible—and highly paid—disc jockeys, who can dominate the personality of the

TABLE 2.1
Radio Stations Formats and Share of Audience[a]

Format	Stations[b] AM	FM	Share of Audience[c] 1989	1987	1985
Contemporary Hit/Top 40	16	297	17%	16%	18%
Urban Contemp/Black	99	94	8	10	9
Album Oriented/Classic Rock	23	249	13	13	11
Adult Contemp/Oldies/Soft Rock	155	424	19	17	16
Jazz/New Age	3	43	2	1	—
Country	193	243	11	11	11
Middle of the Road/Variety	176	5	5	6	7
Spanish Language	75	29	3	3	2
Religion/Gospel	173	64	2	2	2
Classical Music	8	31	2	1	1
Beautiful Music/Easy	24	138	7	8	10
Big Band/Nostalgia	123	9	3	3	4
News/Talk	160	3	10	9	9
Other or Unknown	52	45	1	—	—
Total	1,280	1,674	100%	100%	100%

[a] *Source:* Duncan (1989).
[b] This includes all stations reported in the metro area of measured Arbitron markets. The total of not quite 3,000 stations, of more than 9,000 commerical stations, account for three-quarters of all radio listening.
[c] Spring report for each year. The 1989 share is for the number of stations listed. The 1987 and 1985 share is shown for comparison but would be based on a slightly different number of stations in each category.

station during certain dayparts. This kind of "customized" programming is particularly common in major markets, and it is often accompanied by customized research. Not only do these stations buy and analyze published ratings reports, but they are also likely to make use of computerized access to a ratings database, and engage in a variety of nonratings research projects. We discuss customized analyses of ratings data in chapter 6. The most typical nonratings research includes focus groups, featuring intensive discussions with small groups of listeners, and "call out" research that requires playing a short excerpt of a song over the telephone to gauge listener reactions.

Many stations, however, depend on a syndicated program service or network to define their format. There are dozens of these from which to choose and many are targeted to a specific kind of listener. In the extreme, a station might rely on such pre-packaged material for virtually everything it broadcast, save local advertising and announcements. But in most cases stations do their own "original" programming during the most listened-to morning hours and use a syndicated services—received via satellite or on tape—during the majority of the remaining hours.

In fact, most radio stations do not worry about ratings research either. There are audience estimates for only about half of the nation's 11,000 radio stations are regularly reported by Arbitron. Most stations are in small communities, not major advertising markets. As a result, the majority of all radio time is sold to sponsors who do not have detailed factual ratings information. That is a bit deceiving however, because 3,000 stations that are measured account for more than three quarters of the total radio listening in the country (see Table 2.1), not to mention the lion's share of industry revenues.

We discuss the specific kinds of research questions that can be addressed with ratings data in the pages that follow. But other things being equal, programmers with access to ratings data are in a better position to know their audience than those who have no ratings. Ultimately, that is their greatest value to a programmer. People who work in radio programming know that many of the popular ideas about radio use are not true and can easily be better understood by a careful examination of the ratings book. For example, the 6–10 a.m. daypart, long known as morning drive time, is when radio audiences are largest, but the greatest percent of listeners in automobiles is from 4 p.m. to 6 p.m. Midday, from 10 a.m. to 3 p.m., was long called "housewife time," but in most markets just about as many men are listening to the radio as women. Further, there are often just about as many car listeners during midday than either at morning or afternoon drive time. All in all, the average adult spends about 20 hours a week listening to radio, or nearly 3 hours a day. Contrary to popular belief, teens are not the heaviest users of radio—they listen less than any other demographic group. But teens do comprise the largest part of the audience after 7 p.m. All of these may vary from market to market. As Casey Stengel, or was it Yogi Berra, said: "You could look it up." Assuming, of course, you have the rating report.

Many radio programmers even like to go to Arbitron headquarters to study the diaries that the company has used to collect its ratings data. We will discuss this further in chapter 5. For the moment, all you need to know is that it is a small paper booklet that allows people to make a record of their radio listening. By examining those diaries, a programmer or a consultant working for the station can see if people are remembering call letters or station slogans correctly. Often people write other comments in the diary that are helpful as well.

TELEVISION PROGRAMMING

At the other end of the spectrum is the business of programming a major television network. Although they share some of the same concerns as a programmer in a radio station, they are confronted with different tasks. The most important of these differences is the extent to which a network programmer is involved in the creation of new programs. In that effort, ratings data may be of some value, but as much as anything else, it requires a special talent for anticipating popular trends and tastes, and setting in motion productions that will cater to those tastes. Network programmers like Fred Silverman and Brandon Tartikoff, who demonstrate that talent, can become at least minor celebrities in their own right.

Most other television programmers have less to do with the process of creating programs. Instead, those responsible for programming stations, or even lesser networks, must find their programming elsewhere. Although some original production in the form of news, sports, or collaborative group efforts does take place, most of the programming these people work with is already produced, or already in regular production.

The most common source of programming for stations and networks is what is loosely referred to as the syndication market. In fact, there are many different kinds of syndicated products available to a programmer, and more are being produced everyday. The most obvious source of syndicated programming is material that originally aired on a broadcast network. This material is called *off-network programming*.

Off-network programming is among the most desirable of all syndicated programming. Because it was originally commissioned for a network, it typically has high production values—something that viewers brought up on network fare have come to expect. It also has a track record of drawing audiences, which can be reassuring for the prospective buyer. "M*A*S*H," for instance, was enormously successful on CBS, and has been enormously successful in syndication. In fact, only successful network programs ever make it to syndication. Because syndicated programs are often "stripped" (scheduled 5 days a week), there must be a great many episodes "in the can" to sustain the pace of programming. Although there are a few exceptions, a series must generally have 100 episodes available before it is viable in syndication. That means it must have been on a network for 4 or 5 years, and only the most popular shows last that long.

Because so many program hungry independents and cable networks have appeared in the last decade, the demand for quality off-network product has exceeded the supply. One result, aside from rising prices, is an increase in the number of programs being produced specifically for the syndication marketplace. Traditional "first-run" syndication has included both game shows and talk shows—program types that cost relatively little to produce. Inexpensive or not, shows like "Wheel of Fortune" and the "Oprah Winfrey Show" have been highly successful in the ratings. Less traditional first-run products include "newsier" shows like "Entertainment Tonight" and "PM Magazine."

There are still other sources of programming, like movie packages or regional networks, but whatever their origin, the acquisition of syndicated programming is one of the toughest and most anxiety-producing experiences a TV programmer has to deal with. Usually it involves making a long-term contractual agreement with the distributor, or whoever holds the copyright to the material. For a popular program, that can mean a major commitment of resources. It has been reported, for example, that Liftime, the cable network, paid over $200,000 per episode for off-network reruns of "L.A. Law."

Buying and selling of syndicated programming is often accompanied by an extensive use of ratings data. Distributors will use the ratings to promote their product, demonstrating how well it has done in different markets. The buyers will use the same sort of data to determine how well a show might do in their market, comparing the costs of acquisition against potential revenues.

Once it has been determined what programming a station or network has to work with, television programmers at all levels have more or less the same responsibility. Their programs must be placed in the schedule so as to achieve their greatest effect. Usually that means trying to maximize each program's audience, although some shows are knowingly scheduled against tough competition. For affiliates, the job of program scheduling is less extensive than for others, simply because their network assumes that burden for much of the broadcast day. Even affiliates, however, will devote considerable attention to programming before, during, and just after the early local news. This *early fringe* daypart, just before network prime time, is often the most lucrative for an affiliate. Audience levels are rising and the station need not share the available commercial spots with its network.

As in radio, program production companies, network executives, and stations use a variety of nonratings research to sharpen their programming decisions. This may include the use of one or more measures of program or personality popularity. One company, for example, produces a syndicated research service called TVQ that provides reports on the extent to which the public recognizes and "likes" different personalities and programs. Other program-related research includes theater testing, which involves showing a large group a pilot and recording their professed enjoyment with "voting" devices of some sort. Ultimately, however, ratings are the most important evaluative tool a programmer has. As one network executive said, "Strictly from the network's point of view a good soap opera is one that has a high rating and share, a bad one is one that doesn't" (Converse, 1974).

RESEARCH QUESTIONS

Many of the research questions that a programmer puts to ratings data are, at least superficially, no different than those asked by a person in sales and advertising: How many people are in the audience? Who are they? How often do the same people show up in the audience? This convergence of research questions is hardly surprising because the purpose of programming commercial media is, with some exceptions, to attract audiences that will be sold to advertisers. The programmer's intent in asking these questions, however, is often very different. They are less likely to see the audience as some abstract commodity, and more likely to view ratings as a window on what "their" audiences are doing. These are some of the more common concerns a programmer will have when analyzing ratings data.

Did I Attract the Intended Audience? Because drawing an audience is the objective of a programmer, the most obvious use of ratings is to determine whether the objective has been achieved. In doing so, it is important to have a clear idea of what the intended audience really is. Although any programmer would prefer an audience that is larger than smaller, the often quoted goal of "maximizing the audience" is usually an inadequate expression of a programmer's objectives. More realistically, the goal is maximizing the size of the audience within certain constraints or parameters. The most important constraint has to do with the size of the available audience. That is one reason why programmers are particularly alert to audience shares. Increasingly however, it is not the programmer's intention to draw even a majority of the available audience. In other words, the programmer's success or failure is best judged against the programming strategy being employed.

Probably the clearest instance of declaring intentions is when a programmer goes after a well-defined *target audience*. This is a subset of the audience, usually described in terms of demographics. Virtually every radio format is aimed at a very specific gender age category. For example, one of the currently popular satellite syndicated radio program services by Unistar is "Format 41." It is so named because it is "aimed" at women with an average age of 41. Stations using this syndicated format do not, of course, expect to get only listeners who are 41, however, it does indicate that they are aiming at an audience that is predominantly in the 35–54 age category.

Radio programming in particular seems to lend itself to targeting. An experienced programmer can fine tune a station's demographics with remarkable accuracy. Much of this has to do with the predictable appeal that certain kinds of music will have for listeners of different ages and genders. Table 2.2 lists some of the more common station formats and the extent to which different age groups apportion their listening across formats.

Obviously, you can not expect to attract many youngsters with "big

TABLE 2.2
Share of Audience by Demographic Categories[a]

Age:	12–17		18–24		25–34		35–49		50–64		65+		TOT
Format Gender:	W	M	W	M	W	M	W	M	W	M	W	M	12+
Contemporary Hit	45	36	27	20	17	14	10	8	3	2	1	1	14
Urban Contmp/ Black	25	25	19	16	15	12	10	9	5	4	3	2	12
Album Oriented Rock	13	23	17	35	10	23	3	5	1	1	—	1	11
Adult Contemporary	5	4	11	8	16	13	15	15	8	7	9	5	21
Country	2	3	4	4	5	5	9	11	8	10	5	5	6
Spanish	1	1	3	2	4	3	5	3	3	3	2	2	3
Religion	—	—	1	—	2	1	2	1	2	1	2	1	1
Classical	—	—	—	1	1	2	3	4	4	5	4	5	3
Beautiful Music	1	1	2	1	4	3	12	9	20	18	18	18	9
Big Band/Nostalgia	—	—	—	—	1	1	3	2	10	12	9	12	4
News/Talk	1	2	2	3	5	7	12	15	26	26	42	39	15

[a]*Source:* Arbitron, *Radio Year-round,* 1987. Based on average quarter hour, stations in 14 markets, average of four reports for 1985. Analysis by Walrus Research (George Bailey, David Giovannoni, and Tom Church).

Columns do not add to 100% because of listening to other or unknown stations.

RATINGS ANALYSIS IN PROGRAMMING 33

band" music, or many oldsters with contemporary hits. In chapter 8 we discuss how to depict these sorts of audience characteristics on a "demographic map" of stations that some programmers find quite useful.

TV programmers, too, may devote their entire program service to attracting a particular demographic. This is most evident in some of the new cable networks that have emerged in the last decade. As noted in the last chapter, many networks, like MTV or VH–1, have been programmed to draw certain age groups that their owners believe will be attractive to advertisers. These services are not catering to everyone. In fact, MTV once promoted itself with the slogan "MTV—Some people just don't get it."

Even more conventional TV stations that offer a variety of program types must gauge the size and composition of program audiences against the programming strategies they employ. One commonprogramming strategy is called *counter programming*. This occurs when a station or network schedules a program that has a markedly different appeal than the programs offered by its major competitors. Probably the best example of counter programming is the tendency of independents to show light entertainment (e.g., situation comedies) when the affiliates in the market are broadcasting their local news. The independents are not trying to appeal to the typical news viewer, who tends to be older, and their ratings should be evaluated accordingly.

We should also point out, that it is not only commercial media that care about attracting an audience. Public broadcasting in the United States, or Britain's BBC for that matter, must ultimately justify its existence by serving an audience. If no one is listening, that is very hard to do. Therefore, many public stations use ratings as well. National Public Radio (NPR), with the Corporation for Public Broadcasting (CPB), has provided audience estimates to NPR stations since 1979. The Public Broadcasting Service (PBS), which provides much of our noncommercial television, subscribes to national and local ratings to judge the attractiveness of its programming.

Although they do not have sponsors in the traditional sense, public broadcasters care very much about reaching, even maximizing, their audiences. For one thing, many organizations that put up the money for programming are interested in who sees both their programs and the announcement of who put up the money. The more viewers there are, the happier the funding agency is. Further, public stations are heavily dependent on donations from the audience. Only those who are in the audience will hear the solicitation. Thus, many public TV broadcasters pay considerable attention to their cume ratings.

Even if one has no funding concerns in mind, programmers in a public station well might ask a question like "how can I get maximum exposure for my documentary?" Often, documentaries and how-to-do-it information programs will get nearly the same ratings—very occasionally an even higher rating—when repeated during the daytime, or later at night, and on weekend days than during the first run in prime time. A careful analysis of audience ratings data could reveal when the largest number of those in the target audience are available to view, or whether the intended audience really saw the program.

How Loyal Is My Audience? Audience loyalty is difficult to define precisely because it means different things to different people. We offer a more formal discussion of "channel loyalty" in chapter 9. Informally, loyalty implies the extent to which audience members stick with, or return to, a particular station, network, or program. It is something that manifests itself over time. Despite all the positive images that "loyalty" connotes, we should point out that audience loyalty is quite different from audience size—the attribute most valued by time buyers. Indeed, it is common to hear people talk about a "small, but loyal audience."

Programmers are interested in audience loyalty for a number of reasons. First, in the most general sense, it can give them a better feel for their audience and how they use the programming that is offered to them. That knowledge can be used to guide other scheduling decisions. Second, audience loyalty is closely related to advertising concepts like reach and frequency, so it may have an impact on how the audience is sold. Finally, it can provide an important clue about how to build and maintain the audience you do have, often through a more effective use of promos.

Radio programmers use a number of simple manipulations of ratings data to assess audience loyalty. Although the heaviest radio listening is in the morning when people wake up and prepare for the day, listeners are turning their radios on and off at several times during the day. They also listen in their cars or at work. To maintain ratings levels, a radio station must get people to tune in as often as possible and to listen for as long as possible. Radio programmers employ two related measures, *time spent listening* (TSL) and *turnover,* to monitor this behavior. Using a simple formula based on the average ratings and cume ratings that are in the radio book, one can compute TSL for any station in any daypart. Turnover is the ratio of cume audience to average audience, which is basically the reciprocal of TSL.

If you listen to just about any radio station you can hear how they try to keep you tuned in—naming the records or other items coming up, running contests, and playing a certain number of songs without commercial interruption. To programmers, these tricks of the trade are for "quarter hour maintenance"—that is, trying to keep listeners tuned in from one quarter hour to the next. The 15-minute period is important, because it is the basic unit of time used to compute and report all ratings. By tracking TSL and audience turnover measures, the programmer can see how well the audience is being retained.

Another measure of loyalty that is common in radio programming is called *recycling*. Because ratings books routinely report station audiences for morning and afternoon drive time combined, it is possible to determine how many people listened at both times. This can be a useful insight. If the number is relatively small, for example, it may suggest that programming can be repeated without losing too many listeners.

The same basic research question is relevant in TV programming as well. Do the people who watch the early evening news on a particular station return to watch the late news? If the answer is no, especially if the early news is successful, it would make sense to promote the later newscast with the early news audience. Unfortunately, because of the way TV ratings are published, it is not possible to deduce this from information on the printed page. Customized breakouts of the ratings data will, however, answer that question, and many more.

What Other Stations or Programs Does My Audience Use? In some ways, this is just the opposite of the preceding question. Although there are some people who will listen to one, and only one, station, it is more common for audience members to change from one station to another. In radio, we noted that no two stations are programmed precisely alike or reach exactly the same audience, but several stations in a large market may have very similar, or complimentary, formats. Programmers know that many of their listeners hear four or five other stations in a week. Radio listeners may have favorite times for choosing different formats (e.g., news in the morning or jazz at night). It is important for a programmer to be able to assess the use of other stations.

The use of a ratings book and a few tabulations will enable a programmer to know all the other stations with which he or she shares listeners. Two different sections of the book have the relevant information. *Exclusive cumes,* the number of people who listened to just one particular station during specific dayparts, is the first. Although this is actually a measure of station loyalty, when it is compared with the total cume, it reveals size of the station's audience that has also used the competition. That does not tell you which stations they are, however. A section on *cume duplication* will. It reveals the extent to which the audience for one station is also tuning to each of the other stations in the market. Table 2.3 illustrates levels of audience duplication by stations of different formats.

It will be noted that stations with the largest audiences—biggest shares—usually share with fewer stations. Conversely, those stations with narrower formats, for example jazz/new age in this table, have listeners who make substantial use of different stations and formats. That there

TABLE 2.3
Cume Duplication[a]

Stations by Format[b]	CHR	Urb	CIR	Sof	Old	JNA	Cou	MOR	EZB	BBN	Tlk
CHR	—	41	41	41	27	32	17	19	18	8	14
Urb	26	—	17	14	9	20	6	5	3	2	7
CIR	26	17	—	29	24	30	16	10	3	2	5
Sof	17	9	19	—	13	19	8	11	12	5	5
Old	7	4	10	8	—	5	11	6	9	6	8
JNA	8	8	12	12	5	—	2	5	8	5	5
Cou	9	5	14	11	23	5	—	15	19	13	15
MOR	19	8	16	27	22	20	27	—	47	58	54
EZB	9	2	2	14	18	15	17	23	—	25	20
BBN	4	2	2	6	11	10	12	29	25	—	29
Tlk	6	4	3	4	13	8	11	21	16	22	—
AQH Share[c]	10	6	6	4	4	2	7	12	6	7	4

[a] Figures show the Metro Cume Duplication for 11 radio stations in one market, listed here by their format rather than actual call letters. Read down each column, so that for all listeners who hear the country station any time during the week; 17% also listen to the contemporary hit, and beautiful/easy listening stations, 27% also tune in MOR, etc.
Source: Arbitron (1989)
[b] Formats are: Contemporary Hit Radio, Urban/Black, Classic Rock, Soft Rock, Oldies Rock, Jazz/New Age, Country, Middle of the Road, Easy Listening/Beautiful music, Big Band/Nostalgia, and Talk.
[c] Share, rounded to a whole number.

is so much overlap among stations suggests that people do have varied tastes. They may very well want and enjoy more news, or different kinds of music at different times of the day. Or perhaps people desire a change after listening to one format for a while.

It is clear, however, that stations share the most listeners with other stations of similar appeals. Although they are not all shown in Table 2.3, there are three stations in the market that program mostly "album-oriented rock"—classic rock, soft rock, and a harder "ear bleed station." The average overlap among the three stations is 35%. All told, the number of stations in the market, and especially the number with similar formats, makes a considerable difference in the amount of overlap or cume duplication.

Typically, people listen to two or three different stations during a week. Only 1 in 10 persons listens to only a single station during the week. You can check that for any station or demographic category by looking at the "exclusive cume" section of the ratings book. Most listen to two or three different stations—younger people use more stations than do older listeners, and more different stations are chosen on average in larger markets where more stations are available. You can compute the average number

RATINGS ANALYSIS IN PROGRAMMING

of stations heard for any demographic category in any daypart by adding the "cumes" of all stations and dividing the total by the market cume reported at the bottom of the page.

In large markets, a station may be one of 60 or more radio signals available to listeners. However, the really important competition for most programmers are the other stations trying to reach a similar target audience. These are likely to be stations with a similar format. In general, advertisers will only buy one or two stations "deep" to reach the specific demographic target they seek most. Knowing as precisely as possible where your listeners spend the rest of their radio time is very important.

To get a really detailed look at how listeners tune to different stations, programmers can order a *mechanical diary*. This is a computer printout of all diaries that mention the station. For each person in the sample who heard the station at least once, it provides the exact times they started and stopped listening. Because it lists all diaries that mentioned the station, it also reveals all the other stations these people heard during the week. It is possible, therefore, to see whether your cume audience consists of listeners who used it as their primary station, or heard it only occasionally.

Further, a mechanical diary gives demographic information about each reported listener and the zip code in which they live. The zip code data is often used by stations to find "holes" in their signal area where for technical reasons they may not be heard well. It can also help in buying station advertising. Most stations spend a big part of their advertising budget on billboards and bus cards. Checking the zip codes for present listeners, or for areas where you think you should be reaching more, can help in placing these, as well as other advertising.

Nor is this kind of information valuable only to radio stations. Television offers viewers an increasing number of choices. In fact, the average TV viewer undoubtedly watches more channels than the average listener uses stations. Programmers can use promos for the station most effectively by knowing when different kinds of viewers are tuned in. Sometimes that will mean paying attention to the geo-demographics of the audience, just like an advertiser. But it is especially important to know when people who watch a competitor's program are watching your station. That can be the perfect opportunity to entice those viewers with promotional messages. Nielsen Media Research, in fact, has a computer program called Viewer Tracking Analysis designed to identify just such occasions.

How Will Scheduling Affect a Program Audience?   One of the recurring questions a television programmer must grapple with is how to schedule a particular program. Often, scheduling factors are considered at the time a program is acquired. In fact, some programs sold in barter syndication require stations to broadcast them at a particular time. That is because the syndicators are also selling time to advertisers, and only certain scheduling arrangements will allow them to deliver the desired audience. In any event, how and when a program is scheduled will have a considerable impact on who sees it.

Programmers rely on a number of different "theories" for guidance on how to schedule their shows. Unfortunately, there are nearly as many of these theories as there are programmers. A few of these notions have been, or could be, systematically investigated through analyses of ratings data. Among the more familiar programming strategies are the following.

A *lead-in* strategy is the most common, and the most thoroughly researched. Basically, this theory of programming stipulates that the program that precedes, or leads-in to another show will have an important impact on the second show's audience. If the first program has a large rating, the second show will benefit from that. Conversely, if the first show has low ratings, it will handicap the second. This relationship exists because the same viewers tend to stay tuned, allowing the second show to inherit the audience. In fact, this feature of audience behavior is sometimes called an *inheritance effect*. We discuss factors that enhance or diminish inheritance effects in chapter 7.

Two strategies that also depend on inheritance effects are hammocking and tent-poling. As the title suggests, *hammocking* is a technique for improving the ratings of a relatively weak, or untried, show by "slinging" it between two strong programs. In principle, the second show enjoys the lead-in of the first, with an additional inducement for viewers to stay tuned for the third program. The latter is a kind of "lead-out" effect, if you will. Tent-poling is a less powerful strategy in which one strong show is scheduled between two weak ones. Presumably the first show will enjoy a boost from viewers tuning in in anticipation of the second, and the third will enjoy a conventional lead-in advantage.

Block programming is yet another technique for inheriting audiences from one program to the next. In *block programming,* several programs of the same general type are scheduled in sequence. The theory is that if the viewers like one program of a type, they might stay tuned to watch a second, third, or fourth such program. A variation on block programming is to gradually change program type as the composition of the available audience changes. For example, a station might begin in mid-afternoon when school lets out by targeting young children with cartoons. As more and more adults enter the audience, programming gradually shifts to shows more likely appeal to grownups, thereby making a more suitable lead-in to local news.

All of the strategies described here attempt to exploit or fine-tune how audiences "flow" across programs. Incidentally, many of these programming principles were recognized soon after the first ratings were compiled in the early 1930s. Then, as now, analyses of ratings data allow the programmer to investigate the success or failure of the strategies. Basically, this means tracking audience members over time. Conceptually, the analytical techniques needed to do that are just the same as those used to study the loyalty, or "disloyalty" of a station's audience. We discuss both the theory and practice of such "cumulative" analyses in the last section of the book.

When Will a Program's Costs Exceed its Benefits?   Ultimately, programming decisions must be based on the financial resources of the media. Although some new stations or networks can be expected to operate at a loss during the start-up phases of operation, in the long run the cost of programming must not exceed the revenues that it generates. This hard economic reality enters into a programmer's thinking when new programming is being acquired or when existing programming must be canceled. Because ratings have a large impact on determining the revenues a program can earn, they are important tools in working through the costs and benefits of a programming decision.

When stations assess the feasibility of a new TV program, either a syndicated program or producing a local show, it is typical to start with the ratings for the program currently in that time period and then, based on current ratings and station rates, calculate the revenue that can be generated in that time period. This can be a bit involved, because many factors will affect the revenues a program generates. There is, of course, some uncertainty about the size and composition of the new program's audience. The size of the commercial inventory in the program must be taken into account. If it is bartered, some avails will be gone. Even so, the station may not completely sell out what inventory it does have. The station must also anticipate commissions, for agencies and sales reps, coming out of program revenues. And there are larger marketplace issues, like the strength of the economy, and changes in competing media. All in all, making such projections is a tricky business.

One aide to making these programming decisions, at least when a well-established syndicated program is at issue, is the show's track record in other markets. Often, the local program director can find a market with comparable attributes (e.g., competing programs, lead-ins, etc.) where the show in question is actually airing. There are a group of television ratings reports created for tracking and comparing syndicated programs around the country. In fact, many station reps and program syndicators have these data on computers that allow programmers to quickly explore the consequences of different programming situations.

If a programmer is not agonizing over which programs to acquire, he or she may have to worry about when a program should be canceled. Much of the "press" about television ratings has been over network decisions to cancel specific programs that are well liked by small, often vocal, segments of the audience. Ordinarily, a program will be canceled when its revenue-generating potential is exceeded by costs of acquisition or production. Table 2.4 shows, over the years, the average costs of network prime-time programming, median program ratings, and a "threshold" for cancelation or renewal.

As you can see, the cost of 1 hour of prime time programming has risen steadily over the years. Today, 1 hour of prime-time drama can easily cost over $1 million to produce. On the other hand, the cancellation threshold has fallen over the years. This has happened because the cost of commercial time has increased, as has the total size of the television viewing population. Even so, a prime-time network program with a rating under 14 is not likely to last for long.

The job of programming is probably more challenging today than at any time in the past. There is certainly more competition among electronic media than there has ever been, making the task of building and maintaining an audience more difficult. TV programmers, in particular, must contend with more stations, more networks, and newer technologies (e.g., VCRs, remote controls) that allow viewers to flip, zip, and zap their way

TABLE 2.4
Network Prime-Time Programs: Costs, Ratings, and Cancellations[a]

Year	Average Cost Per Hour	Median Rating Renewed Series	Cancel/Renewal Threshold[b]
1971–1972	$200,279	21.5	17.0–17.1
1972–1973	$205,679	20.0	15.5–18.4
1973–1974	$212,583	21.4	18.3
1974–1975	$212,838	22.2	17.1–18.9
1975–1976	$254,756	21.2	17.7
1976–1977	$310,540	20.2	17.3–18.0
1977–1978	$362,763	20.4	18.3–19.0
1978–1979	$413,100	21.3	17.5–19.1
1979–1980	$418,254	20.8	17.1
1980–1981	$556,102	19.9	16.0–17.5
1981–1982	$571,597	18.4	15.2–16.6
1982–1983	$638,740	18.4	15.2–18.4
1983–1984	$661,058	17.2	15.1–15.5
1984–1985	$725,151	17.1	11.2–14.2
1985–1986	$756,018	17.7	13.8–14.8

[a] Adapted from Atkin and Litman (1986).
[b] All programs with a rating less than the lower threshold were cancelled whereas all programs with a rating above the upper threshold were renewed.

through programming at the touch of a button. In all of these challenges, the analysis of ratings data is likely to offer programmers a useful tool for understanding the audience and its use of media.

RELATED READINGS

Blum, R. A., & Lindheim, R. D. (1987). *Primetime: Network television programming.* Boston: Focal Press.
Eastman, S. T., Head, S. W., & Klein, S. (1989). *Broadcast/cable programming: Strategies and practices* (3rd ed.). Belmont, CA: Wadsworth.
Ettema, J. S., & Whitney, C. D. (Eds.) (1982). *Individuals in mass media organizations: Creativity and constraint.* Beverly Hills: Sage.
Fletcher, J. E. (Ed.). (1981). *Handbook of radio and TV broadcasting: Research procedures in audience, program and revenues.* New York: Van Nostrand Reinhold.
Fletcher, J. E. (1987). *Music and program research.* Washington, DC: National Association of Broadcasters.
Gitlin, T. (1983). *Inside prime time.* New York: Pantheon Books.
Howard, H. H., & Keivman, M. S. (1983). *Radio and TV programming.* New York: Macmillan.
MacFarland, D.T. (1990). *Contemporary radio programming strategies.* Hillsdale, NJ: Lawrence Erlbaum Associates.
Newcomb, H., & Alley, R. S. (1983). *The producer's medium.* New York: Oxford University Press.

RATINGS ANALYSIS IN SOCIAL SCIENCE   3

The electronic media play a central role in the economic and social life of our society. They contribute to the smooth functioning of markets by facilitating the exchange of goods and services. They open or foreclose the "marketplace of ideas" so essential to democracies. Many commentators even argue that they shape our perceptions of reality. Given the powers commonly attributed to the media, it should come as no surprise that they have been scrutinized by social scientists from a wide variety of disciplines. Ratings data have played an important, if little appreciated, role in social scientific inquiry, and in shaping the study of communications.

From the very beginning of radio broadcasting in the 1920s proponents and critics of the medium wondered how it might affect American society. By the early 1930s, a high powered government committee on social trends appointed by President Hoover listed more that 150 specific effects attributable to radio, from homogenizing regional cultures to encouraging morning exercises (Ogburn, 1933). The newly formed networks had also begun assessments of the radio audience, and academics—especially from psychology, sociology, marketing, and education—became interested in the study of broadcasting as well.

In psychology there were already a number of studies comparing the effects of visual versus aural media, thus a comparison between radio and print advertising was almost inevitable. One of the first studies of memory from "ear and eye" was by Frank Stanton (1934), a pioneer in communications research, who would later become the president of CBS. The interest of psychologists broadened considerably and quickly. Stimulated by the use of media for political purposes, especially in the U.S. and Germany, they began to examine the use of radio by President Roosevelt, various religious and political demagogues, and the manipulation of motion pictures by Adolf Hitler. In 1935 Hadley Cantril and Gordon Allport, of Harvard University, published *The Psychology of Radio,* reporting many of their early findings.

By then, Stanton had earned his doctorate from Ohio State, with a dissertation that focused on methods for studying radio listening behavior. He and Cantril sought, and eventually secured, a grant from the Rockefeller Foundation to study the methodologies of measuring radio. As luck would have it, Stanton had become director of research at CBS, and so was unable to head the project. Instead, they asked Paul Lazarsfeld to be the director, with Stanton and Cantril serving as co-directors. Thus began the Princeton Radio Research Project which, after 2 years, moved to Columbia University as the Bureau of Applied Social Research.

Under the auspices of the Bureau came many collaborative research efforts and a string of studies that can be regarded as the beginning of communications research in the United States. Virtually all of this research was tied to the measurement of radio listening. The exciting new field of radio, and especially the need to measure its audience, was largely responsible for establishing Lazarsfeld as a "founding father" of the scientific study of communications. Indeed, the emergence of ratings research, at least in those early days prior to World War II, was intertwined with the development of the new field of mass communications, and in a broader sense, with the growth of all social/behavioral research. Today, the use of ratings data in the social sciences falls into one of two relatively applied areas; financial and economic analysis and communications policy.

FINANCIAL AND ECONOMIC ANALYSIS

It should be obvious by now that audiences are a valuable commodity. They are a critical component in the media's ability to make money, and frequently determine whether a particular media operator succeeds or fails. As the principle index of the audience commodity, ratings data are often used in financial planning and analysis, as well as in more broadly based studies of industry economics. In these applications, ratings information is employed to answer questions that are somewhat different than the ones posed by either an advertiser or programmer.

Those most immediately concerned with the financial implications of ratings data are the owners and managers of the media. Advertiser-supported media are in the business to make a profit. In order to do that, they try to minimize their expenses while maximizing their revenues. We have already seen how this can effect programming decisions. Expenses for the electronic media, however, can also include salaries, servicing debt the firm has incurred, and a host of mundane budget items. One way to improve profits, of course, is to reduce those expenses, but there is a limit to how much cost cutting can be done. The only other way to improve profitability is to increase revenues.

Broadcast stations generate virtually all of their revenues from time sales to advertisers. In radio, the vast majority of revenues come from local advertisers. Television stations, especially those in large markets, get roughly equal amounts of revenue from local and national spot markets. Networks, too, heavily depend on advertising revenues, although cable networks typically derive additional income through direct payments from cable systems. Even program syndicators, who may charge stations for the use of their programming, can realize substantial revenues from selling time to advertisers.

Financial analysts outside the media are also concerned with the profitability of media properties. This category of ratings user includes investors and their representatives. For example, many media companies are "publicly traded," meaning that individual or institutional investors can go to a stock exchange and buy shares in the company. Just as investors would study the prospects of any potential acquisition, a thorough financial analysis is likely to include an inspection of a company's ratings performance—past, present, and future. Even if shares in a media company are not traded on exchanges, investors can buy properties directly. Stations are brokered much like houses. Here again, investors must determine whether the property to be acquired will generate sufficient revenues to make the acquisition worthwhile. Projecting audience ratings is a critical element in making those judgments.

A third category of ratings users are professional economists. Many people with extensive training in economics are involved in the kinds of analyses described here. Economists have also used audience ratings to help shape communications policy, a process we discuss in more detail in the section that follows. In addition to these activities, economists have studied media industries in their own right, just has they might analyze other aspects of the economy. Most economists in the United States are termed neoclassical economists. This is a large and diverse group, who typically study how individuals—through owning, buying, and selling property—contribute to the operation of a marketplace. Some of their research has focused specifically on the buying and selling of audiences. Even political economists, a much smaller group influenced by the works of Karl Marx, have become interested in how the audience commodity is produced, and the role of ratings in that process.

Research Questions

As you might imagine, ratings data have a variety of applications in financial and economic analyses. Despite a good deal of variation in the reasons why researchers ask these questions, most analyses in this area are attempting to answer one of three basic questions.

What Determines the Value of an Audience? We have noted, under a system of advertiser-supported media, that audiences are really a commodity. They are bought and sold just like other commodities. They are "perishable," of course, and their supply is not totally predictable, but that hardly distinguishes them from other goods in the marketplace. Just like other commodities, analysts have tried to figure out what determines their value, at least as it is reflected in prices. Knowing the determinants of a commodity's price is certainly of practical value to those who do the buying and selling, but it can also help us understand the operation of media industries.

The economic value of an audience is largely determined by supply and demand. Corporations and other organizations demand advertising time and the media supply it. Generally speaking, when the U.S. economy is strong, and corporate profits are high, demand increases and advertising expenditures rise. For instance, between 1948 and 1981, there was a remarkably high (.975) correlation between corporate profits and advertising expenditures (Vogel, 1986, p. 174). While such macroeconomic variables establish an overall framework for prices, a number of factors operate within that framework to determine the value of specific audiences.

On the demand side, some companies cannot curtail their advertising expenditures as easily as others. For instance, the makers of many nondurable goods, like soft drinks, cosmetics, and fast foods, fear significant losses in market share if they stop advertising. Consequently, they may continue to advertise heavily, even if times are hard. Local merchants, on the other hand, will quite often cut advertising budgets to reduce expenses. For these reasons, during an economic downturn, local advertising markets may "soften" more readily than national markets, driving down the price of local audiences.

We have already noted that different advertisers demand different sorts of audiences, and that this interest in market segmentation has had a marked effect on the ratings. Audiences are routinely categorized by their demographic, and geographic attributes. Increasingly, they are segmented by psychographics and product-purchasing behavior. Not all audience segments, however, are as easily supplied as others. Some kinds of people spend more time in the audience, and are therefore more readily available to advertisers. Other kinds of people constitute a tiny part of the population (e.g., executives earning more than $500,000) and are, therefore, rare. This tends to make them a more valuable commodity.

All these aspects of supply and demand come into play when determining the value of an audience. Ultimately, such factors are represented in cost calculations (e.g., CPMs) for the electronic media. In fact, advertisers will sometimes make tradeoffs between print and electronic media based on the relative cost of audiences. Table 3.1 summarizes recent CPMs for the major advertiser-supported media. Although such contrasts can be an "apples and oranges" comparison, the price of competing media is another factor that determines the market value of a television or radio audience. This is especially true in local advertising where newspapers can provide stiff competition for the electronic media.

Political economists conceptualize the value of audiences somewhat differently. Here, the central question is one of exploitation. Audience members are, evidently, producing something of real value, just as they would by laboring in a factory. But how does this happen? Is it legitimate to define watching television as "work"? If audience members are working, are they receiving fair compensation? If the media are able to extract some sort of value from the audience without compensation, is the mass audience being "exploited"? Such reasoning is receiving greater attention in Marxist analyses of the media, and to the extent that ratings measure the output of the audience, they may come to play as important a role in the work of political economists, has they have in more traditional economic analyses.

TABLE 3.1
CPM Trends in Major Advertising Media[a]

Medium	1980	1981	1982	1983	1984	1985	1986	1987	1988
Network prime time (HH)	4.94	5.84	6.44	6.94	8.08	8.06	8.78	10.13	9.46
Network daytime (HH)	2.22	2.47	2.68	2.70	3.41	3.48	3.19	3.11	2.82
National spot prime time (HH)	7.35	8.15	8.87	8.15	9.40	10.01	9.57	10.01	11.58
National spot daytime (HH)	2.21	2.63	3.09	2.63	3.03	3.32	3.44	3.77	4.05
Network radio[b]	1.61	1.69	1.81	1.93	2.07	2.22	2.41	2.55	2.65
Spot radio[b]	2.90	3.04	3.28	3.50	3.74	3.86	4.01	4.37	4.52
Consumer magazines[c]	7.34	7.75	8.49	9.18	9.68	10.32	10.92	11.38	11.79
Daily newspapers[d]	14.46	15.76	17.46	19.13	341.33	383.17	402.39	399.19	N/A
Syndicated supplements[c]	6.91	7.81	8.62	10.00	10.36	11.58	9.22	9.65	11.03

[a] *Source:* Burnett (1989)
[b] CPM adults
[c] CPM-Circulation
[d] Inch Cost/MM Circ. Conversion from line to inch rates in 1984. Data are not comparable to previous years.

What Contribution do Ratings Make to Revenues? The preceding discussion runs the risk of suggesting that audiences have some inherent value that translates directly into revenues. A number of factors can account for a discrepancy between audience ratings and audience revenues, and may be of considerable importance to both economists and financial analysts.

The first thing to remember is that audiences of the electronic media are, themselves, invisible. The only index of that commodity is ratings data. In a very real sense it is ratings points that are bought and sold. As long as such audience estimates are the only way to know the size and shape of the commodity being traded, they effectively become that commodity. Athough ratings companies are under considerable pressure to produce accurate audience measurements, certain biases and limitations do exist. Some may be inherent in the research methods these companies use, others are more the result of how the ratings business itself has responded to the demands of the marketplace. In any event, the media must generally operate within the constraints imposed by the ratings, and that may be a hindrance to selling certain audiences.

For example, we have noted that cable has gradually eroded the audience for broadcast television. The cable industry, however, has had some difficulty marketing that audience because the ratings business has been geared to estimating broadcast audiences. Only recently have ratings companies begun to use methods well suited for measuring cable networks. Even the format of a ratings report can affect an advertiser's willingness to use the data. We detail the evolution, methods, and products of the ratings services in the second section of the book, but for now, recognize that ratings data themselves can distort the link between audiences and audience revenues.

The second thing to remember is that audiences are made available to advertisers in the form of spot announcements. These spots, however, are limited in number. A broadcaster could, therefore, exhaust the inventory of available spots before meeting the demand for audiences. The result being that some audience revenues would go unrealized. The amount of advertising time that the electronic media has to sell is affected by several things. By tradition, certain dayparts have more commercials than other dayparts. Prime time, for instance, has fewer spot announcements than late night or daytime television. Network affiliates have less time to sell to local advertisers than independents, because network programming reduces the size of their inventories. Inventories can be increased by adding commercial time to a program or reducing the duration of spots (e.g., from 30 to 15 seconds), but like cost cutting, there is a practical limit to how much can be done without being counterproductive. Indeed, radio broadcasters frequently argue about how much commercial time can be sold within each hour before listeners might be driven away to another station. Stations frequently try to lure listeners to a new, or revised, format by presenting very few commercials or guaranteeing "X" commercial-free minutes.

Even if ratings data were completely accurate and inventories more flexible, audiences would not necessarily determine revenues. A sales force must take that audience data into the marketplace, and persuade advertisers to buy the commodity. Selling is a very human, and often imperfect, process. Some sales managers are more aggressive than others in their approach to time buyers. Some salespeople are more effective in dealing with clients than others. The net result is that two audiences that seem to be identical may sell for different amounts of money. In fact, advertisers who buy several spots at once, routinely receive quantity discounts.

Economists have devoted a good deal of attention to the relationship between audience ratings and audience revenues. In addition to the factors we have already described, there are other, less benign, explanations for a discrepancy between an audience's size and its market value. If, for instance, there are relatively few competitors in a market, they may be tempted to collude and set prices above competitive levels. Although we know of no cases of such collusion, the potential exists. What is clear, however, is that the demand does dictate price. Advertiser demand for TV has been, and will probably remain, high. This, as best as can be determined, has meant higher rates (proportional to the audience delivered) in markets with fewer stations. In less concentrated markets, the cost of audiences may be lower. Studies of this sort have not been conclusive. For readers who are interested in learning more about the relationship between the audiences an media revenues, Table 3.2 lists research in the area.

As we have said, such research has not, and may not, provide conclusive answers, but in recent years we note the growing number of independent TV stations. These outlets have more available commercial time, per segment, than network affiliates. This increase on the "supply side," with many other choices—cable and VCRs—has meant a slow growth in rates in recent years.

Financial analysts also recognize that although audiences are an important determinant of media revenues, there may well be some discrepancy between a media property's share of the audience and its share of market revenues. This can have practical implications for evaluating the desirability of different acquisitions. Table 3.3 illustrates how a financial analyst might go about evaluating the long-term revenue potential of a television station.

RATINGS ANALYSIS IN SOCIAL SCIENCE

TABLE 3.2
Determinants of Audience Revenues

Potential Determinants	References [a]
Audience size	Fisher, McGowan, & Evans (1980)
	Fournier & Martin (1983)
	Fratrik (1989)
	Levin (1980)
	Wirth & Bloch (1985)
Audience demographic composition	Fisher et al. (1980)
	Fournier & Martin (1983)
	Fratrik (1989)
Audience location (MRA/ADI/TSA)	Fisher, et al. (1980)
	Fratrik (1989)
	Poltrack (1983)
Certainty of audience delivery	Fournier & Martin (1983)
Daypart	Fisher et al. (1980)
	Poltrack (1983)
Overall strength of market economy	Poltrack (1983)
	Vogel (1986)
Number of stations in the market	Fournier & Martin (1983)
	Levin (1980)
	Poltrack (1983)
Number and circulation of newspapers	Poltrack (1983)
Market power or concentration	Fournier & Martin (1983)
	Wirth & Bloch (1985)
Ratio of nat'l spot to local revenue	Poltrack (1983)
Level of cable penetration	Wirth & Block (1985)
VHF versus UHF	Fisher et al. (1980)
	Fratrik (1989)
	Levin (1980)
Total levels of media use in market	Poltrack (1983)
Season of the year	Poltrack (1983)
Size of sales transaction	Fournier & Martin (1983)

[a] References do not necessarily find the same relationship between determinants and audience revenues.

The top line across the table represents the net market revenue for all television stations in the market. This number is likely to be a function of the overall market economy, especially the annual volume of retail sales. It is estimated by looking at historical trends in the market, and making some carefully considered judgments about the economic outlook for those sectors of the economy that are especially important in the market. The second line represents the station's current and estimated share of the television audience. Here again, the analyst would consider recent trends and the chances that the station's overall ratings performance will improve or decline. A more specific discussion of factors affecting a station's ability to attract an audience is provided in our model of audience behavior in chapter 7. In this particular example, the analyst

TABLE 3.3
Station Revenues Based on Audience Share Projections

CH	1989	1990	1991	Maturity
Net market revenue	$70 million	$74 million	$80 million	×
Station audience share	21%	22%	23%	25%
Over/under sell factor	.80	.83	.86	.90
Station revenue share	16.8%	18.3%	19.8%	22.5%
Station Revenue	$11,760 million	$13,540 million	$15,840 million	×(.225)

estimated that the station would eventually be able to attract and hold 25% of the audience.

That does not necessarily mean that the station can expect to capture 25% of total market revenues. In fact, this station has regularly gotten a smaller share of market revenues than its share of the audience. In other words, it "undersells" its audience share. That factor is recognized in the third line across the table. The analyst believed that the undersell factor could be improved, but to be conservative, projected that share of revenue would always fall short of audience share.

Once these factors have been estimated, it is possible to make a reasonable projection of station revenues. When these revenue estimates are compared with projected operating expenses, the analyst can determine whether this property would have sufficient cash flow to cover its debt, and provide the owners with an acceptable return on their investments.

What Economic Factors Affect Audience Ratings? Because ratings can be a powerful determinant of media revenues, much attention is devoted to understanding what will affect the size of audience ratings. Some of these factors have already been mentioned in our discussion of programming, and others are dealt with in the chapters that follow. But, a few efforts to examine the impact of economic factors deserve brief mention here.

One question that often confronts media operators is whether a particular expenditure will increase audience ratings. This might involve something as pedestrian as an engineering expense to improve the quality of a transmission. Frequently, it involves a decision about how much money to spend on programming. Clearly, there is not a perfect correlation between a program's cost and its ratings performance. Some very expensive programs fail, while shows with relatively modest budgets attract substantial audiences. Overall, however, there appears to be a significant relationship between the amount of money spent on programming, and the size of the audience that programming ultimately attracts. Some economists, for example, have argued that because the English-speaking marketplace is relatively large, producers of English-language programming are willing to invest more in production, resulting in films and television programs with greater audience appeal (e.g., Wildman & Siwek, 1988).

At a different level of analysis, researchers have considered the impact of economic factors like media ownership on ratings performance. A great many broadcast stations are owned by corporations that have other media properties. These might include other media outlets in the same market, or additional stations across the country. It is not unreasonable to expect that such media groups would enjoy certain "economies of scale" in the acquisition or production of programming. That is, they could spend more on salaries and facilities, because they are able to distribute those costs over the entire media group. They might also have greater leverage in dealing with program syndicators. As a result, group ownership is an economic factor that we would hypothesize is positively related to ratings performance. In fact, there is evidence that this is the case.

It should be noted, however, that the advantages of group ownership merit continued scrutiny. During the past several years distributors have forced more stations into competitive bidding for syndicated programming. This is, in large part, because of the growth of independent stations. In the 1980s the number of such stations devoted to "general entertainment"—that excludes religious, Hispanic, and shopping channel outlets—increased from only 100 to about 265. This has had the effect of strengthening the hand of syndicators in their negotiations with stations—a factor that may reduce the advantages of group ownership.

COMMUNICATION POLICY

It is in the making of communications policy where one sees the broadest range of interests represented, as well as the most inventive uses of ratings data. Most of the players involved in the process are intent on securing some special advantage for themselves, or somehow handicapping their opponents. Others seem genuinely interested in promoting what they believe to be in the "public interest." No matter their intent, most have found some occasion to use ratings data to help make their case.

This regular, and sometimes unthinking, use of ratings information has a number of causes. Most significantly, it is that audience ratings go to the heart of the media's power. Why is it that electronic media have any economic value at all? It is because they have an audience. Why is it that news and entertainment programs are capable of any social impact? It is because they have an audience. Although ratings information, alone, cannot reveal the effects of the media on society, they can frequently index the potential. These factors, in addition to the wide and continuous availability of ratings data, have made ratings an attractive tool in crafting communications law and regulation. Before considering the research questions that ratings data can address, however, we need to briefly review the people and institutions most involved in making communications policy. It is these players, and the dynamic interactions among them that determine the course of public policy. Those most likely to use ratings in developing policy are the federal government, industry, and the public.

By the mid-1920s it became apparent that broadcasting could not be left to operate as an unregulated marketplace. It appeared that more people wanted to broadcast than there were frequencies to accommodate them. To solve the problem of who was to broadcast on which frequencies, the U.S. Congress created the Federal Radio Commission, which was eventually replaced, in 1934, by the Federal Communications Commission (FCC). The FCC was to license stations, and more generally, see to it that the "public interest" was served in the process. Although prevailing philosophies of what best serves the public interest have changed over the years, three interrelated objectives have endured. First, the commission has tried to limit certain undesirable social effects that could be attributed to broadcasting. Many times, children have been the special object of their concern. Second, the commission has tried to promote greater diversity in media content, often by structuring markets in a way that makes them more responsive to audience demand. Third, like many other regulatory agencies, the FCC has tried to ensure the overall economic health of the industry it regulates.

The last of these is surely the subject of the most argument. Many broadcasters would point out that the FCC has in fact acted to hinder "free, over-the-air" TV by allowing—indeed openly promoting—a more rapid growth of cable in recent years. Although it is true that the commission has tended to promote competition in the electronic media, it has historically stopped short of policies that would inflict a debilitating measure of "economic injury" upon licensees. For example, this concern is evident in the commission's approach to questions of audience "diversion" which we review later.

In any event, the FCC does not have a free hand to implement whatever policies it chooses. Other institutions within the federal government are often involved. The president, the courts, and especially the Congress, can and do make their wills known from time to time. Other independent agencies, like the Federal Trade Commission or the Copyright Royalty Tribunal, will deal with matters of communications policy. And other elements of the executive branch, like the Departments of Commerce or Health and Human Services may enter the picture. For example, in the early 1970s the Surgeon General oversaw a massive study of the impact of violence on television. Further, although the government ultimately sets communications law, other interests weigh heavily in the policy-making process

The entities with the most direct interest are the media themselves. Sometimes individual companies, like the broadcast networks, represent themselves in Washington. Usually however, industries are represented by trade associations. For broadcasters, the most important of these is the National Association of Broadcasters (NAB). The NAB serves the interests of commercial broadcasters by lobbying Congress and the FCC, testifying before Congressional committees, filing briefs in the relevant judicial proceedings, and participating in rulemakings and inquiries at government agencies. In many of these activities, the NAB will submit research that bears on the issue at hand. In fact, the NAB has a special department of research and planning, that frequently performs policy studies using ratings data. The cable industry is represented by the National Cable Television Association (NCTA). The NCTA engages in the same sorts of activities as the NAB and likewise maintains a department of research and policy analysis. The other trade associations that are most likely to use ratings data include the Motion Picture Association of America (MPAA), and the Independent Television Association (INTV).

Even the combination of government and industry does not completely control the formation of public policy. The public itself enters the process in a number ways. Most directly, of course, we elect government representatives. Occasionally, one of these government officials will take the lead on a matter of communications policy, inviting us to either support or reject that position. More organized public participation comes in the form of "public interest groups." Some of these, like Action for Children's Television (ACT), are formed specifically to affect communications policy. ACT, in particular, has been successful at drawing the attention of Congress and the FCC to matters of children's television. Other groups, like the National Congress of Parents and Teachers (PTA) or the American Medical Association (AMA), do not make communications law and regulation their central focus, but nonetheless express occasional interest in social control of media.

We should also note that the academic community contributes to policymaking. Professors in those disciplines with an enduring interest in broadcasting and the electronic media have been attracted to policy questions dealing with the media's social and economic impact. The academic community can affect policy in several ways. Most notably, they do so by publishing research relevant to questions of public policy. Because they are often viewed as experts, and relatively objective, their work may carry some special influence with the government. They may also work as consultants for other players in the process, and so exercise influence in a less public, but equally direct manner. Finally, of course, they can indirectly affect policy through their students. Many people who are, today, involved in determining communications policy, would credit certain professors with influencing their views on matters of law, regulation, and social responsibility.

Research Questions

Not all, or even most, questions of communications policy can be illuminated by analyses of ratings data. But there is a surprisingly broad range of applications for ratings analysis. In fact, very few bodies of social scientific data can be interpreted, or "read," in so many different ways. The use of ratings data in policy-making can be organized as responses to one of three broad questions. These questions correspond to those three long-term concerns of the FCC: (a) limiting the undesirable effects of media, (b) promoting more diverse and responsive programming, and (c) tending to the economic condition of its client industries.

What do the Media do to People?   This is commonly known as the "effects question." We have been asking it, in one form or another, since the earliest days of mass communication. Even before radio was established a number of sociologists and educators were concerned about the impact of crime reported in newspapers on children. In the early 1930s sociologists tried to determine the impact that movies had on young people. Later in the decade psychologists studied the effects of wartime propaganda, while marketing researchers measured how press coverage could influence voter behavior. More recently, we have concerned ourselves with television's role in promoting violence, sexual or racial stereotypes, and distorted perceptions of social reality.

Central to these, and all other effects questions, is some "cause and effect" relationship. In its general form, the question is, "Does exposure to the media (cause), make other things happen (effect)?" These are extremely difficult questions for social scientists to answer. An important starting place, however, is a knowledge of what people are exposed to. This is because any direct media effect must begin with exposure.

Although an encounter with the media may not determine a particular outcome (again, the effect), it defines a certain potential. If many people use a medium, or see an item of content, the potential for effects is great. Conversely, if no one is exposed to a message, its impact is never felt. Advertisers have long realized this, and so have paid dearly for access to audiences.

Academics, too, have recognized that exposure is the wellspring of media effects. One of the most outspoken has been George Gerbner, a proponent of "cultivation analysis." Gerbner has argued that television content is so uniform, and people are so unselective, that researchers need only consider the amount viewed in order to determine the medium's social impact. Usually, these arguments are buttressed with references to ratings information.

> According to the 1984 Nielsen Report, the television set in the typical home is in use for about 7 hrs a day, and actual viewing by persons older than 2 years averages over 4 hrs a day. With that much viewing, there can be little selectivity. And the more people watch, the less selective they can and tend to be. Most regular and heavy viewers watch more of everything. (Gerbner, Morgan, & Signorielli, 1986, p. 19)

According to Gerbner, then, TV's power to cultivate mistaken notions of what is real can be revealed in simple comparisons of heavy and light viewers.

Other academic researchers are less convinced that audience selectivity is a moot point. In fact, studies of "selective exposure" have an important place in the history of effects research. Although varied in their origins, these studies assume that audience members are capable of a good deal of "choosiness," and will demonstrate that in their consumption of media content. Depending on the kinds of content chosen, different media effects may follow. In the Surgeon General's report on TV violence, for example, Israel and Robinson (1972), used viewing diaries to assess how much violence was consumed by various segments of the population. The operating assumption was that those who consumed more violence-laden programming would be more likely to show its ill effects. So, whether one considers specific content, or, as Gerbner would argue, TV viewing in general, it is clear that exposure sets the stage for subsequent media effects.

Government regulators have also used audience information to assess the media's potential to create socially undesirable effects. The FCC, for instance, has a Congressional mandate to limit the use of indecent language in broadcasting. Although the commission might tolerate certain excesses if they were heard only by adults, the presence of children in the broadcast audience has created a problem. So much so, that the commission has tried to "channel" offensive language away from those time periods when there is a "reasonable risk" that children will be in the audience. To identify those time periods, the commission's staff has used ratings data. Hence, the detrimental effects that might result from exposure to indecent content are, at least, limited in scope by the size of the child audience.

In a similar vein, the FCC has expressed special concern about the audience for local news and public affairs programming. The commission has a long history of encouraging "localism" in broadcasting. This effort has been motivated, at least in part, by a desire to keep people informed about issues of public importance. The growth of cable television has been seen as a threat to localism because of its ability to divert audiences from local broadcasts. In a 1979 report on the relationship between cable and broadcasting, the commission elaborated,

> Television may have an important effect in shaping the attitudes and values of citizens, in making the electorate more informed and responsible, and in contributing to greater understanding and respect among different racial and ethnic groups... Historically, the FCC has encouraged particular types of programming—local news, public affairs, instructional programs—on these grounds. To the extent that a change in broadcast-cable policy would dramatically change the amount by which these programs are not only broadcast but viewed, these issues could be an important component of a policy debate. (Federal Communications Commission, 1979, p. 639)

In a line of reasoning analogous to its indecency rules, then, the FCC expressed concern that undesirable social consequences might flow from people *not* watching certain content. Here again, the first index of effects is the size of the audience, an index that is readily available in ratings data.

What do People Want? Another important goal of communications policy has been to provide the public with diverse media content. This objective is very much in keeping with our First Amendment ideals, and the benefits that are thought to result from a "free marketplace of ideas." But how does one accomplish this objective? Although policymakers have different opinions on that subject, the most popular solution has been to structure media industries so that a large number of firms can compete for the attention of the audience. In such an environment, it is believed, competitors will probe audience demand, responding to their likes and dislikes as expressed in their program choices.

Under this system, ratings can be thought of as a kind of feedback mechanism. Arthur Nielsen, Jr. (1988) has described the link between ratings and preferences as follows:

> Since what the broadcaster has to sell is an audience to advertisers, it follows that in order to attract viewers, the broadcaster must cater to the public tastes and preferences. Ratings reveal these preferences. (Nielsen, 1988, p. 62)

Many commentators find the industry's argument that they only "give the people what they want," to be self-serving and deceptive. Most media, they would point out, are responding to the demands of advertisers, not audience members. Because some audiences are less valuable to advertisers than others, these viewers may be underserved. Additionally, the use of advertiser-supported media does not provide audience members with a way to express how much they like a particular program, only that they have elected to use it. We review a number of other factors that complicate the link between preference and choice in chapter 7. Nevertheless, a considerable body of theory, in both psychology and economics, views people's choices as a function of their preferences, and this provides more than adequate justification for the use of ratings in policymaking.

The most relevant of these theories has been developed in the study of *welfare economics,* a branch of the discipline that is concerned with how we can maximize the "welfare," or overall well-being of society. Like other branches of economics, it assumes that people are rational beings who will attempt to satisfy their preferences when choosing goods and services. At least, that is, insofar as their pocketbooks allow. Economists refer to this notion as the "theory of revealed preference." Indeed, they make a case that deducing preferences from behavior may be superior to direct questions about a person's likes and dislikes. A media system under advertiser support, however, imposes no direct costs on viewers (i.e., they do not pay a per program fee). Viewer preferences, therefore, can be freely expressed in program choices. These concepts, and their consequences for how public policy might maximize viewer satisfaction, are fully discussed in Owen, Beebe, and Manning (1974).

Welfare economists, therefore, have used ratings data to address questions of communications policy. One category of FCC rules that has received scrutiny is the commission's ownership rules. In an effort to increase diversity in programming, the commission has sought to limit certain classes of media from owning local television stations (e.g., local newspapers and radio). The idea is that different owners will contribute different viewpoints to the marketplace of ideas. Unfortunately, existing media may be more adept at offering local programming that appeals to viewer preferences. Parkman (1982) has, consequently, argued the following:

> If these classes of owners produce more popular programming than other classes of owners, the reduction in popular programming should be taken into consideration as cost of the diversification policy. To determine if certain news gathering organizations are more successful than others in attracting viewers, we can look at the end result that these organizations produce as judged by the viewers, i.e., the ratings. (pp. 289–290)

After analyzing the ratings of local television news programs, Parkman concluded that the commission's policy imposed, "costs on individual viewers by forcing them to choose programs considered by them as less desirable" (p. 295).

The FCC has, itself, used ratings as a kind of revealed preference. The most notable example has been the commission's designation of stations as "significantly viewed." This concept was introduced into FCC rules in the early 1970s. Cable systems, which at that time almost exclusively depended on retransmitting broadcast signals to attract subscribers, faced strict limits on the distant signals they could "import." If, however, it could be demonstrated that a signal from outside the market was, nonetheless, significantly viewed, then the system was allowed to carry it. In fact, they could be compelled to carry it as a *must-carry* signal.

In the original rules, a station was deemed to be *significantly viewed* if, in noncable households, it achieved a weekly 2% share of audience, and 5% weekly circulation. The FCC actually used Arbitron data to list all the stations that were significantly viewed in each U.S. county. Although the commission's motivation for imposing this standard is not entirely clear, it seems that this was one attempt to let the viewers themselves demonstrate that a station provided a valued local service.

As the number of specialized cable networks began to grow, the cable industry increasingly viewed must-carry stations as an annoyance that prevented them from offering more desirable cable only services. In 1985, the courts agreed, and judged the commission's rule to be unconstitutional. By that time, however, the concept of a significantly viewed station had worked its way into other bodies of law.

Under copyright law, a cable system may carry a local station by paying an inexpensive, blanket "compulsory license." Tests of significant viewership are used to help determine which stations are local, and may be extremely important to stations that would not otherwise be carried. The Cable Policy Act of 1984, requires that there be "effective competition" in a cable system's market area before it can be freed of rate regulation. Tests of significant viewership are used to determine the number of competitors, and may be extremely important to cable systems that would not otherwise be free to set rates.

One gets a sense of the special place that ratings data hold in policymaking, by considering just how a finding of significant viewership is made. Arbitron produces special "County Coverage" reports from the viewing diaries it collects each year. These are reported for cable and noncable households. Audience information is reported for each subset of the audience. However, because so few diaries are collected in some counties, share and circulation estimates are sometimes based on as few as 10 diaries. As we explain in chapter 5, that leaves a lot of room for error. Most of the people who use the data are perfectly aware of that problem, but barring a specially commissioned survey, that is the standard that everyone has agreed to. Ratings data, even that modest, are used because they offer an "objective," readily available, benchmark.

What Economic Implications Will Various Policies Have? A great many government laws and regulations affect the financial condition of the media and related industries. Because these policies have an impact on the "bread and butter"—or in some cases the Mercedes and BMWs—of those businesses, they attract the attention of many participants in the policymaking process. Even the FCC, which in recent years has favored increased competition in the media, must remain alert to the economic consequences of various policies. After all, the commission is responsible for seeing to it that broadcasting serves the public interest. If broadcasters are driven out of business by some ill-conceived government policy, it might compromise the commission's mandate.

Financial statements that describe the media's revenues, expenses, and profitability, are one obvious source of information on the economic condition of the industry. But for a number of reasons, these data are not always used. For one thing, the commission several years ago stopped collecting financial statements from broadcasters. For another, economic injury to the industry might be too far advanced by the time it shows up on company ledgers. One common alternative to a dollars-and-cents measure of economic impact is to use audience ratings. Because ratings measure the commodity that the media have to sell, policies that adversely affect a station's audience are often seen to damage its economic interests. Despite the differences we noted between ratings and revenues, evidence of lost audiences is often effective evidence of lost revenues.

Most illustrative of this use of ratings information is a succession of studies that have attempted to demonstrate audience "diversion" or "erosion" from established media. Such analyses have been a frequent feature in skirmishes between broadcasters and the cable industry. Almost from the beginning, broadcast interests used claims of "economic injury" to encourage policies that would restrict cable's growth. Allowing cable to enter a market, it was argued, would so erode a station's audience as to threaten its very survival. In 1970, Rolla Park, of the Rand Corporation, assessed this threat through an analysis of local market ratings data. This study helped shape the FCC's rules on cable television, issued in 1972. The commission again considered the economic relationship between cable and broadcasting in an inquiry in the late 1970s. Again, Park (1979) and a number of interested parties, assessed the state of audience diversion through rather sophisticated analyses of audience ratings information.

Again, the commission made extensive references to these studies in its final report.

More recently, the FCC encountered claims of audience diversion in the context of its rules on *syndicated exclusivity*. These rules, first adopted in the early 1970s, were intended to insure that broadcasters who bought the exclusive rights to syndicated programming, would not have that privilege undermined by a cable system that imported a distant signal with the same program. The import, it was assumed, would divert some audience that rightly belonged to the local station. Although the commission latter dropped the rule, in 1988 the FCC decided to reimpose syndicated exclusivity. During that proceeding, the parties at interest (e.g., NAB, NCTA, INTV) all submitted analyses of ratings data purporting to show that audience losses did or did not occur in the absence of the rule. In reimposing the rule, the commission reasoned that

> the ability to limit diversion means broadcasters will be able to attract larger audiences, making them more attractive to advertisers, thereby enabling them to obtain more and better programming for their viewers. ("The Why's," 1988, p. 58)

The economic implications of ratings data have also had a substantial impact on the operation of the Copyright Royalty Tribunal (CRT). The *compulsory license* fees paid by cable systems for the right to carry broadcast signals create an annual pool of money in excess of $200 million. The CRT is responsible for determining how that money should be distributed among the various claimants. Those with a claim are copyright holders including program suppliers, commercial broadcasters, public broadcasters, and Canadian broadcasters. The largest group, program suppliers, is represented by the Motion Picture Association of America (MPAA).

To distribute compulsory license fees, the CRT uses a two phase process. In the first phase, the CRT allocates varying percentages of the total pool to the major categories of claimants. To make a case that it deserves the largest piece of the pie, each year MPAA commissions a special study from A.C. Nielsen, that is based on ratings data from Nielsen's Station Index. Nielsen provides MPAA with a list of every qualifying program, the number of quarter hours it was broadcast, and the number of cable households that viewed each quarter hour. The MPAA, in turn, argues that the copyright pool should to be allocated in accordance with the amount of viewing attributable to each group of copyright holders. Although CRT distributions deviate somewhat from this scheme, MPAA have been successful in winning about 70% of the total pool using this strategy.

In the second phase of the proceedings, the representatives of each category of claimants distribute their share of the pie to individual copyright holders. The MPAA makes those allocations in strict adherence to the Nielsen data. Although this method of distribution is not without error, it seems to be one of the few techniques that virtually all participants can agree to. Here again, the ratings seem to offer an objective, unbiased standard. Further, in this particular application, their use seems a reasonable way to determine the revenues that are otherwise lost to copyright holders. After all, the economic value of a syndicated program rests largely on its ability to attract an audience.

The uses of ratings data in legal or regulatory proceedings is considerable. Despite these, and many other applications of the data, it appears to us that social scientists have only scratched the surface of the analytical possibilities. For the most part, these uses of the ratings have dealt with gross measures of audience size. Perhaps that should not be surprising. Such estimates are the most readily available. Indeed, that is what "the ratings" are. But as we stated at the outset, this book attempts to explore all the ways in which the ratings database can be exploited. Using ratings to track individuals over time, engaging in what we call *cumulative analyses*, would seem a logical next step for social scientific inquiry.

Take, for example, the effects question. Although the size of the audience exposed to a message may suggest something about its potential effect, so too does the regularity of exposure. Advertisers have recognized this concept in their attention to frequency of exposure. Effects researchers might similarly ask how often people see or hear a particular kind of programming. Do all children see about the same amount of violence on television, or do some consume especially heavy doses? Is there a segment of the child audience who seem to be violence junkies? If so, who are those children? Do they come from poor or affluent families? Do they watch alone or with others? The answer to such questions, all of which can be gleaned from ratings data, might contribute much to our understanding of the impact of televised violence. Similar questions could be asked about the audience for news and information.

Studies of audience duplication might reveal more about people's preferences for programming as well. Does a particular program have a "small-but-loyal" following, or is it just small? Programmers and marketing researchers have long recognized a certain feature of audience duplication called "channel loyalty." Religious, Spanish language, and music video services are among the kinds of programming that seem to attract relatively small-but-loyal audiences. Does this intensity of use suggest something about how the audience values a service above and beyond the number who use it at any point in time?

The economic value of an audience is also affected by things other than its size and composition. We have seen that advertisers may specify reach and frequency objectives in their media plans. Those who seek a high frequency of exposure might be willing to pay a premium for that small-but-loyal audience. In a similar vein, it would be interesting to explore how channel loyalty and inheritance effects contribute to the audience of a syndicated program. If a station "adds value" to the program by delivering an audience predisposed to watch, then perhaps the station should have a greater share of compulsory license fees.

Even media critics who are often distrustful of social scientific methods might learn more about how the audience members encounter the media through inventive uses of ratings data. For instance, analysts of popular culture have become increasingly interested in how people "read" or make sense of television programming. One insight from this line of research is that the viewers experience the medium not as discrete programs, but as strips of textual material called *flow texts*. It might be illuminating to explore the emergence of flow texts through analogous studies of audience flow.

All of these analyses, and many more, could be realized through the application of ratings data. Unfortunately, the effective use of ratings data in the social sciences and related disciplines has been uneven. In part, that is because the proprietary nature of modern ratings makes them too expensive for strictly "academic" analyses. We have more to say about buying data in chapter 6. Some academics, however, may fail to exploit the data that are available, simply because they do not recognize the possibilities for analysis. We hope the remainder of this book helps remedy the latter problem. The following sections acquaint the reader with the ratings services, the data they collect, the products they offer, as well as the theory and techniques of ratings analysis.

RELATED READINGS

Besen, S. M., Krattenmaker, T. G., Metzger, A. R., & Woodbury, J. R. (1984). *Misregulating television: Network dominance and the FCC*. Chicago: University of Chicago Press.

Bryant, J., & Zillmann, D. (Eds.). (1986). *Perspectives on media effects*. Hillsdale, NJ: Lawrence Erlbaum Associates.

Hogarth, R. M., & Reder, M. W. (Eds.). (1986). *Rational choice: The contrast between economics and psychology*. Chicago: University of Chicago Press.

LeDuc, D. R. (1973). *Cable television and the FCC: A crisis in media control*. Philadelphia: Temple University Press.

LeDuc, D. R. (1987). *Beyond broadcasting: Patterns in policy and law*. New York: Longman.

Levin, H. J. (1980). *Fact and fancy in television regulation*. New York: Russell Sage.

Lowery, S., & DeFleur, M. L. (1987). *Milestones in mass communication research: Media effects* (2nd ed.). New York: Longman.

Noam, E. M. (Ed.). (1985). *Video media competition: Regulation, economics, and technology.* New York: Columbia University Press.

Owen, B. M., Beebe, J. H., & Manning, W. G. (1974). *Television economics.* Lexington: Lexington Books.

Rowland, W. D. (1983). *The politics of TV violence: Policy uses of communication research.* Beverly Hills: Sage.

Vogel, H. L. (1986). *Entertainment industry economics: A guide for financial analysis.* Cambridge: Cambridge University Press.

PART II

RATINGS DATA

THE RATINGS RESEARCH BUSINESS 4

Since the beginning of radio, the broadcaster has been interested in how the owner of a receiver reacts to the programs presented over the air. Some of the questions to which the broadcaster, whether he is an educator or advertiser, is anxious to secure the answers are as follows:

1. *When does the listener use his receiver?*
2. *For how long a period does he use it?*
3. *To what station or stations does he listen?*
4. *Who listens (sex, age, economic and educational level)?*
5. *What does he do while the receiver is in operation?*
6. *What does he do as a result of the program?*
7. *What are his program preferences?*

—Frank N. Stanton (1935)

Surprisingly little has changed since Stanton wrote those words. The electronic media, themselves, have undergone tremendous change, but the basic research question—a need to know the audience—has been one of the most enduring features of the industry. In this chapter we trace the evolution of the ratings business. Our purpose is not to offer a comprehensive history of audience measurement. For the reader who wants such detail, we recommend Beville's (1988) book on the history and methods of ratings research. Our interest in the growth of ratings is motivated by a desire to better understand the industry's present condition, and perhaps to anticipate its future.

Even the first "broadcaster" wanted to know who was listening. After more than 5 years of research, experimentation, and building on the work of many others, Reginald A. Fessenden broadcast the sound of human voices on Christmas Eve in 1906. He played the violin, sang, recited poetry, and played a phonograph record. Then, the electrical engineer promised to be back on the air again for New Year's Eve and asked anyone who had heard the broadcast to write him. Apparently, he got a number of letters especially from radio operators astonished to hear more than Morse code on their headphones. Other early station operators asked for letters from listeners as well. In fact, Dr. Frank Conrad, who in 1920 developed KDKA in Pittsburgh for Westinghouse, played specific records requested by his correspondents.

A need to know the audience, however, quickly became more than just a question of satisfying the operator's curiosity about unseen listeners. By the early 1920s AT&T had demonstrated that charging clients a toll to make announcements over the station was an effective way to fund the medium. It was only a short step from the concept of "toll broadcasting" to the notion of selling commercial time to advertisers.

By 1928, broadcasting was sufficiently advanced to provide listeners with consistent and quality reception. Many people had developed the habit of listening to radio, and broadcasters in cooperation with advertisers were developing program formats "suitable for sponsorship" (Spauling, 1963). Although there was some public controversy over whether radio should be used for advertising, the Great Depression, beginning in 1929, caused radio station owners to turn increasingly to advertisers for support.

For radio to be successful as an advertising medium, however, time buyers had to know who was in the radio audience. Newspapers were already providing authenticated figures on circulation through the Audit Bureau of Circulation. Theater and movie audiences could be measured by ticket sales. Phonograph popularity could be measured by sales and, later, juke box plays. But broadcasters, and their advertisers, were left with irregular, and frequently inadequate, assessments of the size and composition of the audience.

Many radio advertisers, for example, offered coupons or prizes in an attempt to measure response. Indeed, in 1933 about two thirds of all NBC advertisers made some sort of offer that listeners could send for—mostly novelty items, booklets with information, or a chance to win a contest. Sometimes, responses were overwhelming. In answer to a single announcement on a children's program, WLW in Cincinnati, got more than 20,000 letters. The program's sponsor, Hires Root Beer, used these responses to select specific stations on which to advertise in the future. But soliciting listener response had risks. The makers of Ovaltine, a drink for children, and the sponsors of Little Orphan Annie, asked fans to send in labels in order to free Annie from kidnappers. As you might imagine, this provoked an uproar from parents.

Early stations used equally primitive techniques to estimate the size of their audience. Some counted fan mail, others simply reported the population, or number of receivers sold, in their market. Each of these methods was unreliable and invited exaggeration. The networks were somewhat more deliberate about audience measurement. In 1927, NBC commissioned a study to determine not only the size of its audience, but hours and days of listening. It also sought information on the economic status of listeners, foreshadowing the use of "demographics" now so much a part of audience research. In 1930, CBS conducted an on-the-air mail survey, offering a free map to all who would write the station to which they were listening. CBS compared the response to the population of each county, and developed its first "coverage" maps. But none of these attempts offered the kind of regular, independent measurement of the audience that the medium would need to sustain itself.

THE BEGINNING OF RATING RESEARCH

The history of ratings research is a story of individual researchers and entrepreneurs, of struggles for industry acceptance, as well as an account of the broadcasting business itself. It is also a story of research methods. Every major ratings research company rose to prominence by perfecting and promoting its own brand of research. Most major changes in the structure and services of the industry have also been tied to research methods. For this reason, we have chosen to organize our discussion of the industry's history around the research methods these firms have used.

Telephones

From 1930 to 1935 the revenues, and the profits, of the network companies nearly doubled, all at a time when the country and most other businesses were in a deep economic depression. Because many American families did not have money to spend on other diversions—and because radio was indeed entertaining—the audience grew rapidly. An important stimulant to that growth, however, was the emergence of a system for providing audience estimates that advertisers could believe. The first such system depended on another technological marvel—the telephone.

Then, as now, advertisers were the driving force behind the ratings research, and it was advertisers who helped create the first regular ratings company. In 1927, a baking powder company hired the Crossley Business Research Company to survey the effectiveness of its radio advertising. Two years later the company did a similar survey for Eastman Kodak using telephone interviews to ask people if they had heard a specific program. Although the telephone was, at the time, an unconventional tool for conducting survey research, it seemed well suited for measuring something as far-flung and rapidly changing as the radio audience.

Archibald Crossley, the research company president, and a well-known public opinion pollster, suggested to the Association of National Advertisers (ANA) that a new industry association might use the telephone to measure radio listening. His report, entitled "The Advertiser Looks at Radio," was widely distributed and ANA members quickly agreed to pay a monthly fee to support a regular and continuous survey of radio listening. The American Association of Advertising Agencies (AAAA) also agreed on the need for regular radio audience measurements.

This new service, officially called the Cooperative Analysis of Broadcasting, or CAB, began in March 1930. More often than not, however, their reports were referred to in the trade press as the Crossley ratings. Even the popular press began to note the rise or fall of a specific program or personality in the ratings. Initially, only advertisers paid CAB for its service, but soon after, advertising agencies began to subscribe. The networks had access to the reports as well, using them for selling and making programming decisions, but they could not make "official" use of them. Significantly, it was not until 1937 when NBC and CBS were allowed to become subscribers, thus sharing the cost.

Crossley revised his methods and expanded the amount of information he provided a number of times in the early years. By the 1935–1936 season, surveys were being conducted in the 33 cities that had stations carrying CBS and the two NBC networks. Calls were placed four different times during the day and respondents were asked to "recall" the radio listening during the last 3 to 6 hours. Hence, Crossley's method of measurement was known as *telephone recall*. Monthly, and later bi-weekly, reports were provided giving audience estimates for all national network programs. Further, three times a year there were more detailed summaries providing detailed reports on station audiences hour by hour, with breakdowns for geographic and financial categories.

There were, however, problems with the CAB's methods. One problem was measuring radio listeners who did not have telephones. Oddly enough, this was less of a problem in the early years of the service because the first families to purchase radios were higher income households likely to have telephones. But the growth of radio homes soon far out paced those with telephones, and by the end of the 1930s CAB had to alter its sampling procedures to include more low-income homes to compensate.

But the most serious limitation to the CAB method was that it required listeners to recall (remember) what they had heard. Despite the method's efficiency—it could collect information on listening to several hours of programs—another technique that featured a simultaneous or "coincidental" telephone survey was soon to challenge Crossely's early dominance of the ratings business.

George Gallup had measured radio audience size by conducting personal interviews and asking what station they were listening to as early as 1929 at Drake University in Iowa. Soon thereafter, he went to work for Young and Rubicam, a major advertising agency, where he did a *telephone coincidental* on a nationwide basis. There were other pioneers of the telephone coincidental like Pauline Arnold, Percival White, and John Karol who was later director of research for CBS.

In 1933, Pauline Arnold specifically compared the telephone recall and coincidental methods as summarized in Lumley (1934):

> The results showed that some programs, which were listened to by many listeners, were reported the next day by only a few. In general, dramatic programs were better remembered than musical programs. However, the rank correlation between the percentage of listeners hearing 25 (half hour) programs and the percentage reporting having heard them was about .78. This is a measure of the adequacy of the Crossley survey as compared with the simultaneous telephone survey. (pp. 29–30)

The telephone coincidental provided a methodological advantage that opened the door for CAB's first ratings competitor. This happened when Claude Hooper and Montgomery Clark quit the market research organization of Daniel Starch in 1934 to start Clark-Hooper. Dr. George Gallup assisted them in arranging for their first survey. Hooper later wrote that "Even the coincidental method which we have developed into radio's basic source of audience size measurement was originally presented to us by Dr. George Gallup" (Chappell & Hooper, 1944, p. vii). In the Fall of that year, Clark-Hooper launched a syndicated ratings service in 16 cities.

Ironically, Clark-Hooper was first supported by a group of magazine publishers who were unhappy with the fact that radio was claiming an ever larger share of advertiser dollars. They believed that Crossley's recall technique overstated the audience for radio. Although it could be expected then that coincidental ratings would capture certain unremembered listening, the publishers hoped that Clark-Hooper would show that many people were not home, and many others at home were not listening to the radio. In fact, the first Clark-Hooper results did show lower listening levels than those of CAB.

In 1938, Clark-Hooper split, the former taking the company's print research business. With great faith in the future of radio, Hooper went into business for himself. His research method was simple. Those answering the phone were asked:

- Were you listening to the radio just now?
- To what program were you listening?
- Over what station is that program coming?
- What advertiser puts on that program?

Then they were asked to report the number of men, women, and children who were listening when the telephone rang.

Hooperatings, as his audience estimates came to be called, were lower than CAB's for some programs but higher for others. As Hooper would argue later, people were better able to remember programs that were longer, more popular, and those that had been on the air for a longer period of time. Respondents were also much more likely to recall variety programs, and most likely to forget having listened to news (Chappell & Hooper, 1944, pp. 140–150). Over time, the industry began to regard C.E. Hooper's coincidentals as more accurate than CAB's recall techniques.

But methodological superiority was not enough. As the creature of the ANA and AAAA, CAB was well entrenched with the advertising industry. Recognizing that, Hooper decided to pursue the broadcast media themselves, arguing that "Whereas CAB was established to serve the buyer of radio time, the object of C.E. Hooper, Inc. was and is to furnish audience measurements to both the buyer and the seller of radio time." If CAB saw fit to ignore networks and stations, Hooper would seek them out as clients, and provide them with the kinds of audience research they needed. This strategy was perceptive, for today, it is the media who account for the overwhelming majority of ratings service revenues.

Hooper also worked hard for the popular acceptance of Hooperatings. To achieve as much press coverage as possible, each month he released information on the highest rated evening programs. This went not only to the trade press, but to popular columnists as well. In this way, C.E. Hooper, Inc. became the most visible and talked about supplier of audience information for the industry. Radio comedians even began to joke about their, or the competition's, Hooperatings.

In addition to promoting popular consciousness about program ratings, Hooper was also responsible for establishing many of the traditions and practices of contemporary ratings. He instituted the "pocketpiece" format for ratings reports, now the hallmark of the national Nielsen ratings, as well as concepts like the "available audience" and "sets in use." He also began to report audience shares, which he called "percent of listeners," and the composition of the audience in terms of age and gender. Thus, even by the end of the 1930s the basic pattern of commercial audience research for broadcasting was set.

Hooper and his company were efficient and aggressive. He regularly did research to try to make his methods more accurate or to add new services, especially to help the networks and stations. He was also relentlessly critical of the CAB method that still depended on recall. As a part of this battle, in 1941, Hooper hired Columbia University psychology professor Matthew Chappell to study recall and memory. Two years later they wrote a book trumpeting the advantage of telephone coincidental.

Hooper's aggressiveness paid off and just after World War II, he bought out CAB, which was on the verge of collapse. C.E. Hooper was now the unquestioned leader in ratings research. But even as Hooper reached his zenith, broadcasting was changing. The number of radio stations expanded rapidly, and the new medium of television was about to alter the way in which people used their leisure time. A new methodology and company were ascendant as well. Although he continued to offer local measurement of radio and television, in 1950, Hooper sold his national ratings service to A.C. Nielsen.

Personal Interviews

In-person, formal interviews often were used in early radio surveys, especially by academics with a sociology or marketing background. Though the method is sometimes used today to measure Hispanic audiences, for example, it is no longer a mainstay of the ratings industry. Nonetheless, knowing something about the method, and those who used it, offers valuable insights into the impact that ratings can have on the industry it is trying to serve.

Most of the early Daniel Starch studies for NBC, beginning with the first in Spring 1928, were done by personal interview. And even though the first ratings services had come into existence, in the 1930s CBS commissioned Starch to do a series of reports. CBS argued that this provided more accurate information because Hooper's "telephone calls obviously miss all non-telephone homes—which becomes an increasing distortion as one gets into the smaller communities." Because CBS had fewer and, often, less powerful affiliated stations than NBC, the network felt it could only benefit from this sort of audience research (CBS 1937).

In the late 1930s, while Crossley and Hooper argued over different methods of telephone data collection, and Nielsen worked to perfect his metering device, the personal interview was still the most accepted method of collecting socio-psychological behavioral information. One man in particular, Dr. Sydney Roslow, who had a doctorate in psychology, became intrigued with the technique while interviewing visitors at the New York World's Fair. With the encouragement of Paul Lazarsfeld, he started to adapt these techniques to the measurement of radio listening.

In the Fall of 1941, he began providing audience estimates, called *The Pulse of New York,* based on a personal interview "roster recall" method that he developed. When respondents were contacted they were given a list of programs, or roster, to aid in their recall of listening for the past few hours. Because Hooper, and later Nielsen, concentrated on network ratings, his local service expanded rapidly—especially with the tremendous expansion of stations after the War. By the early 1960s Pulse was publishing reports in 250 radio markets around the country and was the dominant source for local radio measurement.

The roster recall method had some significant advantages over its competitors. It could include out-of-home listening (e.g., automobile and work), and measure radio use during hours not covered by the telephone coincidental—Hooper was limited to calls from 8 a.m. to 10:30 p.m. Further, it provided much more demographic details and information on many minority and foreign language stations popular with those less likely to have phones.

The widespread availability of local radio ratings that captured information on listeners who were hard to reach with other methods had a significant impact on the shape of local radio itself. It contributed to the rise of R&B and other contemporary hit music on "Top 40" stations that catered to those listeners. Similarly, because Pulse measured only metropolitan areas it proved an aid to the growth of "formula" rock stations. Those stations—as compared with older, more powerful network affiliates—did not have large coverage areas and were disserved in comparisons of large regional audiences. Top 40 depended on local, not network or national spot, advertisers, and local sponsors were interested in the most popular stations in their service area, not listeners hundreds of miles away. Thus, Pulse was a boon to the growth of rock formats, just as more and more local stations were coming on the air, and more and more network programs and personalities were transferred to TV or oblivion.

However, by the 1970s still another ratings company, featuring another method, took control of local radio ratings. The American Research Bureau (ARB), which we describe in the sections that follow, used its success with television diary techniques to move into radio. As a subsidiary of a large computer company, ARB had superior computing power that aided in the timely production of market reports. It also appears that the rock and ethnic stations favored by the interview method were not as aggressive in selling to advertising agencies, so agencies came increasingly to accept the diary technique being promoted by news and "easy listening" stations. In 1978, Pulse went out of business.

Meters

The advantage of making a permanent, continuous record of what people were actually listening to, as it happened, was obvious from the beginning of radio. But the technical problems to be solved were far from easy, so such systems were not developed until the 1930s, nor were they in common use until the late 1940s. When these meters finally arrived, however, they had a profound and lasting impact on the ratings business.

While a student at Columbia University in 1929, Claude Robinson—later a partner with George Gallup in public opinion research—patented a device to "provide for scientifically measuring the broadcast listener response by making a comparative record of . . . receiving sets . . . tuned over a selected period of time" (Beville, 1988, p. 17). The patent was sold to RCA, parent of NBC, but nothing more is known of the device. Despite the advantages of a meter, none had been perfected, leading Lumley (1934) to report:

> Although the possibilities of measurement using a mechanical or electrical recording device would be unlimited, little development has taken place as yet in this field. Reports have been circulated concerning devices to record the times at which the set is tuned in together with a station identification mark. None of these devices has been used more than experimentally. Stanton, however, has perfected an instrument which will record the exact time at which a radio set is turned on. (pp. 179–180)

The reference was, of course, to Frank N. Stanton, then a graduate student at Ohio State and later the president of CBS. For his dissertation, the first paragraph of which began this chapter, Stanton built and tested 10 recorders "designed to record set operation for periods as long as 6 weeks." On wax-coated tape, one stylus marked 15 minute intervals while another marked when the set was turned on. The device did not record station tuning but was used to check against listening as recorded on questionnaires. Stanton, by the way, found that respondents tended to underestimate the time they spent with the set on, a bias of recall techniques that holds true even today.

In 1930 and 1931 Robert Elder of the Massachusetts Institute of Technology conducted studies of radio's advertising effectiveness that were published by CBS. In 1933–1934, he and Louis F. Woodruff, an electrical engineer, designed and tested a device to record radio tuning. The device scratched a record on paper by causing a stylus to move back and forth as the radio tuner was moved across the dial. Elder called his device an "Audimeter," and sought a patent. Discovering the previous Robinson—now RCA—patent, he received permission from RCA to proceed. The first field test used about 100 of the recorders in the Boston area. In 1936, Arthur C. Nielsen heard a speech by Elder describing the device and apparently began negotiating to buy the rights to the technique immediately.

An electrical engineering graduate of the University of Wisconsin, Nielsen had first opened a business to test the efficiency of industrial equipment. When he began in 1923, it was a period of great expansion for inventing, manufacturing, and the rapid deployment of new assemblyline techniques. The business survived but did not prosper. In 1933, a pharmaceutical client suggested to a Nielsen employee that what they really needed was information on the distribution and turnover of their products. In response, Nielsen developed a consumer survey based on a panel of stores to check inventory in stock. The business grew fast, a food index was added, and the company prospered. The A.C. Nielsen Company was on its way to becoming the largest marketing research firm in the world. But it was the acquisition of the Elder-Woodruff audimeter that would ultimately serve to imprint the Nielsen name in American's consciousness.

With his engineering background, and the profits from his successful indices, Nielsen redesigned the device. There were field tests in Chicago and North Carolina to compare rural listening, in 1938. Despite war shortages, by 1942 Nielsen launched a "radio index" or NRI, based on some 800 homes equipped with his device. Nielsen technicians had to visit each home periodically to change the paper tape in the device, which slowed data collection. However, the company also provided information about product purchases, based on an inventory of each household's "pantry." Having already established a good reputation with advertisers, Nielsen began to make progress on overtaking the dominant ratings supplier, C.E. Hooper.

During the 1950s, Nielsen continued to expand his ratings business and to perfect the technology of audience measurement. As we noted, in 1950 he acquired Hooper's national ratings service. In the same year he initiated the Nielsen Television Index or (NTI), the company's first attempt to measure that fledgling medium. By the middle of the decade, he launched the Nielsen Station Index, or (NSI) to provide local ratings in both radio and television. His engineers perfected a new version of the audimeter that recorded tuner activity on a 16mm film cartridge. More importantly, the cartridge could be mailed directly to Nielsen sample households, thereby speeding the rate at which data could be collected.

Nielsen had also begun to use diaries for gathering audience demographics. To improve their accuracy, he introduced a special device called a recordimeter, that monitored hours of set usage, and flashed a light to remind people to fill in their diaries.

The 1960s were more tumultuous, not just for Nielsen but for all ratings companies. In an atmosphere charged by quiz show scandals on television, reports of corruption and "payola" in the music industry, as well as growing social unrest, the U.S. Congress launched a far-reaching investigation of the ratings business. Recognizing the tremendous impact that ratings had on broadcasters, and concerned about reports of shoddy research, Oren Harris, chairman of the House Committee on Interstate and Foreign Commerce, orchestrated a lengthy study of industry practices. In 1966, the Harris Committee issued its report. Although it stopped short of recommending legislation to regulate audience measurement, the investigation had sobering effects on the ratings business—effects that are still evident today in the scrupulous detail with which methods and the reliability of ratings are reported, and the existence of the Electronic Media Rating Council (formerly the Broadcast Rating Council).

As the premier ratings company, Nielsen was particularly visible in the congressional hearings, especially its radio index. In response, Mr. Nielsen personally developed a new radio index that would be above criticism. Unfortunately, potential customers resisted the change because of the increased costs associated with data collection. Angered by this situation, in 1964 Nielsen withdrew from national radio measurement altogether. In fact, a year earlier Nielsen had discontinued local radio measurement, leaving Pulse unchallenged. To this day, television is the only medium Nielsen measures.

Diaries

In the 1920s, many radio set builders and listeners were not interested in programs at all. Instead, they were trying to hear as many different and distant stations as possible. In order to keep track of those stations, they kept elaborate logs of the signals they heard, when they heard them, and noted things like station call letters, city of origin, slogans, and program titles. Despite this early form of diary keeping, and the occasional use of diaries by radio ratings firms, the "diary" method did not become a mainstay of commercial audience research until the rise of television.

The first systematic research on dairies was done by Garnet Garrison. In 1937 he began to "experiment developing a radio research technique for measurement of listening habits which would be inexpensive and yet fairly reliable" (Garrison, 1939, p. 204). Garrison, for many years a professor at the University of Michigan, noted that at the time the other methods were the telephone survey, either coincidental or unaided recall, personal interviews, mail analysis or surveys, and "the youngster automatic recording." His method, which he called a "listening table," borrowed something from each because it could be sent and retrieved by mail, included a program roster, and was thought to be objective. His form provided a grid from 6 am to midnight divided into 15 minute segments, and asked respondents to list station, programs, and the number of listeners. He concluded that:

> With careful attention to correct sampling, distribution of listening tables, and tabulation of the raw data, the technique of "listening tables" should assist materially in obtaining at small cost quite detailed information about radio listening. (Garrison, 1939, p. 205)

CBS experimented with diaries in the 1940s, but apparently thought of the data as applicable only in programming. It was, therefore, used to track such things as audience composition, listening to lead-in or lead-out programs, and charting audience flow and turnover. In the late 1940s, Hooper also added diaries to his telephone sample in areas "which cannot be reached practically by telephone." This mixture of diary and coincidental was never completely satisfactory. Indeed, one of the reasons for the slippage of Hooper against Nielsen was that the telephone method was confined largely to large metro areas, where TV first began to erode the radio audience. Hence, Hooper tended to understate the radio audience.

It was not until the late 1940s that diaries were introduced as the principle method of a syndicated research service. As director of research for the NBC owned station in Washington, DC, James Seiler had proposed using diaries to measure radio for several years. The station finally agreed to try a survey for its new TV station. NBC helped pay for several tests, but Seiler set up his own company to begin a regular ratings service.

He called the company American Research Bureau (ARB), and in Washington, just after the war, its name sounded very official, even patriotic. ARB issued its first local market report in 1949. Based on a week-long diary, which covered May 11–18, it showed the Ed Sullivan "Toast of the Town" Sunday variety program with a 66.4 rating. "Wrestling," on the ABC affiliate at a different time got a 37.5, and "Meet the Press" on NBC got a 2.5.

By the fall the company was also measuring local TV in Baltimore, Philadelphia, and New York. Chicago and Cleveland were added the next year. The company grew slowly at first—as both TV and the diary method gained acceptance. Diaries were placed in TV homes identified by random phone calls. From the beginning, Seiler was careful to list the number of diaries placed, and those "recovered and usable." Further, "breakdowns of numbers of men, women, and children per set for specific programs [could] be furnished by extra tabulation" (American Research Bureau, 1947, p. 1).

Another research company had begun diary based ratings in Los Angeles in 1947, using the name Tele-Que. The two companies merged in 1951 thus adding reports for Los Angeles, San Diego, and San Francisco and bringing to ARB several young, bright researchers such as Roger Cooper and R. R. "Rip" Ridgeway, who would help lead the company's growth.

Through the 1950s, ARB emerged as the prime contender to Nielsen's local TV audience measurement, especially after 1955 when it took over the local Hooper TV ratings business. The company expanded, and by 1961 it was measuring virtually every TV market twice a year, and larger markets more often. The networks and stations responded by putting on especially attractive programming during these "sweeps" periods. Local radio reports, also compiled from diaries, were begun in 1965. These, as we have seen, eventually put Pulse out of business, and for many years, left ARB the undisputed provider of local radio ratings.

ARB also attempted to one-up Nielsen by developing a meter whose contents could be tapped by a telephone call. In 1957, ARB installed phone lines in 300 New York City households and began to provide day after ratings with an "instantaneous" meter. Generally speaking, this move met with the approval of advertisers and the media because it meant Nielsen might face more effective competition. Unfortunately for ARB, Arthur Nielsen and his engineers had patented almost every conceivable way of metering a set. ARB's new owner, a firm named CEIR, was forced to pay Nielsen a fee for the rights to the device. Nevertheless, this quickly spurred Nielsen to wire a New York sample with meters, and later, in 1973, to introduce a Storage Instantaneous Audimeter (SIA) as the data-collection device for its full national sample.

By 1967, ARB was acquired by a computer company named Control Data Corporation. Smarting from its run in with Nielsen, ARB used the new owner's technical expertise to develop a metering technology which would not infringe on Nielsen patents. In 1973 it also changed its name to Arbitron. What sounded so patriotic after World War II, evoked a "big brother" image after the turbulent 1960s. A name change, it was thought, might improve response among suspicious respondents. The diary, however, remained the backbone of Arbitron's ratings research business.

THE RATINGS BUSINESS TODAY

There are four major suppliers of ratings research in the United States today. Two are relative newcomers to the business—Statistical Research, Inc. and Birch Radio. Much like the companies described earlier their creation and growth are clearly attributable to individual entrepreneurs. The other two, Nielsen and Arbitron, are large companies, owned by even larger corporations. Although leadership in the industry can certainly change over time, it would be fair to say that Nielsen and Arbitron are currently the dominant players.

Nielsen is the sole supplier of national TV network ratings, although for a time in the 1980s it appeared as if a serious contender would challenge for a share of the market. Audits of Great Britain (AGB) had long supplied England and a number of countries in Europe and Asia with ratings research. With a new measurement technology called the peoplemeter, AGB hoped to establish itself in the U.S. market. As we see in the next chapter, peoplemeters expanded the capabilities of traditional household meters by allowing viewers to enter information about who was watching television. AGB worked hard to get funding from the industry, including advertisers and the media. Within a couple of years it had sufficient support to wire the Boston market with peoplemeters, and begin a field test of the system. Nielsen responded by announcing plans to test and to implement a national peoplemeter service of its own. In 1987, Nielsen began basing its NTI services on a sample of households equipped with peoplemeters. AGB held on for a time, but with equivocal support of the industry, especially the broadcast networks, its position was untenable. In 1988, it announced it would end its U.S. operations.

The introduction of peoplemeters reveals a good deal about the ratings business in America. On one hand, AGB received genuine encouragement from the industry, especially advertisers. This was not unlike the support ARB got when it tried to best Nielsen with an instantaneous meter in the late 1950s. Almost everyone, except Nielsen, is inclined to believe that competition would lead to improved services and lower costs to clients. In fact, the AGB threat undoubtedly accelerated the implementation of Nielsen's peoplemeter, although the company had been experimenting with that technology for years.

On the other hand, the introduction of peoplemeters was accompanied by complaints from ratings users ranging from biases in the data, to too much data being served up too fast, to data that was not provided in a useful or usual format. So, despite a desire for innovation, there is also a kind of inertia that grips the people who use the data. Constancy in the supply of ratings data—knowing what is coming from one month to the next, or being able to make comparisons one year to the next—does have its value. Therefore, changes in the production of ratings must be reconciled with the industry practices that have grown up around the supply of certain data.

The introduction of peoplemeters is revealing for another reason. As we see in the following chapter, no ratings research method is perfect. No system of estimating audiences is without certain biases. Occasionally, those biases will operate to the advantage of some, and the disadvantage of others. Because the peoplemeter system does a better job of measuring small, demographically targeted audiences, advertiser-supported cable networks are likely to be beneficiaries. This is one reason why the broadcast networks were a bit cool about introducing peoplemeters in the first place. If peoplemeter data allow cable to compete more effectively with the major networks for advertiser dollars, it might ultimately have an impact on the kinds of programming we see on television. The point is, not only does industry demand shape the nature of ratings data, but the availability of certain kinds of data can shape the industry, too—just as the Pulse ratings encouraged the development of local radio.

In 1984, the A.C. Nielsen Company was sold by the Nielsen family to Dun & Bradstreet, a company that specializes in credit reports and other information services. It is likely that Nielsen will use this new affiliation to expand its consumer marketing services, exploiting wherever possible economies of scale in the collection or integration of data bases.

The Arbitron Company, has continued to grow as well. Unlike Nielsen, however, it now faces competition in all of the ratings markets it serves. As the sole supplier of local radio measurement, Arbitron's radio ratings business had been very lucrative. On one hand, there were many stations and, hence, many potential subscribers. On the other hand, a single ratings survey could collect data for all those buyers. Although advertisers and stations alike would try to negotiate for the lowest possible price, the potential for profits was considerable. So much so, that where Arbitron was once unchallenged, a new competitor emerged.

As a radio programmer, Tom Birch conducted his own research that helped him develop a very popular format. Soon several stations asked him to do "call out" research for them too. From this, he gradually started to measure radio use. By 1980, he was providing service to 18 markets. Today, Birch Radio is providing market reports in over 250 markets nationwide. Although Arbitron has proved difficult to dislodge, Birch has begun to receive a measure of industry acceptance.

Statistical Research Inc. (SRI) was formed in 1969 by Gerald Glasser, a statistics professor at NYU, and Gale Metzger, former director of research for the Nielsen Media division. Three years later, SRI took over operation of a collaborative industry research effort called Radio's All Dimension Audience Report (RADAR®), for which Glasser had been a consultant. RADAR was the industry's attempt to fill the void in radio network ratings left by the demise of Nielsen's NRI. Since that time, SRI has provided bi-annual reports on radio network audiences.

Interestingly, both Birch and SRI have returned to using the research method pioneered by Archibald Crossley—telephone recall. These companies, of course, make extensive use of computers, and long distance calling from centralized facilities. Further, the near universal penetration of the telephone has minimized many of the problems of sample bias inherent in early applications of the method. Although some limitations still exist, the comparative advantages of this technique, which we discuss in the next chapter, make such services quite viable today.

In television, with fewer stations in each market, the number of potential buyers for ratings products is reduced. Although it was once common for television stations to subscribe to two services, increased competition among stations and tightening budgets have led many to choose a "single service" for ratings information. Today, 75% of all TV stations subscribe to Nielsen's NSI, 66% get Arbitron ratings, but only about 50% subscribe to both.

In some markets, one television ratings service is clearly the dominant provider, although both companies measure every market during sweeps, whether they have enough clients to cover the costs or not. Providing meter-based local overnights is another matter, however. The ratings companies try to secure clients before installing this relatively expensive form of measurement. As a result, several good-sized markets are served by only one metered service. The key to winning the majority of business in a market is, just as it was for CAB and Hooper, industry acceptance. Arbitron and Nielsen are fully aware of this, and market their services aggressively, especially to agencies.

As the competition heats up, ratings suppliers will continue to try to improve their "bottom lines" by introducing new services and reports that make use of their data bases. Unfortunately, many prospective buyers are ill equipped to evaluate the increasing flow of ratings information. Indeed, one area in which ad agencies and stations seem willing to cut expenses is in hiring personnel to deal with media research. In the long run, this combination could pose a threat to the integrity and reliability of the audience measurement industry. As Gale Metzger has warned, there are "... too many naive buyers who will take any kind of information and use it because it is there; too many suppliers who will provide data without the first concern for quality, because they are salable" (1984, p. 47).

Assuming that greed and ignorance don't destroy the credibility of the business, ratings data are likely to remain a powerful presence for many years to come. These numbers have been a central feature of the broadcast industry and the public's perception of that industry for over half a century. Networks, stations, advertising agencies, syndicators, and virtually every other related business prosper or suffer by them. One need only glance at the trade press to realize how pervasive ratings data are. In fact, the general public now receives rather detailed ratings reports in publications like *USA Today, The New York Times,* and many other daily newspapers. As the electronic media become more competitive, as advertisers seek increasingly targeted markets, and as the methods of research and analysis become increasingly sophisticated, it seems certain that ratings will continue to influence the shape and the psyche of American media.

RELATED READINGS

Beville, H. M. (1988). *Audience ratings: Radio, television, cable* (rev. ed.). Hillsdale, NJ: Lawrence Erlbaum Associates.

Chappell, M. N., & Hooper, C. E. (1944). *Radio audience measurement.* New York: Stephen Daye.

Lumley, F. H. (1934). *Measurement in radio.* Columbus, OH: The Ohio State University.

RATINGS RESEARCH METHODS 5

A great many decisions are made on the basis of ratings data. We have seen that billions of dollars are spent on the media in accordance with the ratings. Perhaps it is even fair to say that millions of lives are affected by the programming and policy decisions that hinge on ratings information. Yet we have also seen that no method for producing ratings data is without certain biases or limitations. It is, therefore, important for a ratings analyst to understand where the data come from and how research techniques affect the final product.

Although the practice of ratings research has obviously changed over the years, certain issues have endured. For the most part these have involved questions about research methods. Matters of audience sampling and measurement are as important today as they were when Archibald Crossley launched the CAB. They will, undoubtedly, define future debates about the quality of ratings data as well.

This chapter describes the methods now in use by the ratings services. We do not intend to review every technical detail in the production of ratings data. That would be more than most readers want or need. For those who wish a timely and detailed "description of methodology," the ratings firms will provide the necessary documents. It is our intention to give readers enough of a grounding in the research methods these companies use so that they can assess the strengths and weaknesses of various ratings products, and understand most of the technical jargon they will encounter in ratings reports. We begin with a discussion of sampling.

SAMPLING

All ratings data estimate what is occurring in a population. All ratings companies do this by studying a subset of the population called a *sample*. This strategy is employed because most populations are too big to be studied in their entirety. It would simply take too long and cost too much money to contact everyone. That is certainly the case with a population of over 90 million television households. Virtually all survey research from marketing studies to public opinion polls depends on sampling. Indeed, sampling is used in many scientific endeavors. As Arthur Nielsen, Sr. was fond of saying "if you don't believe in sampling, the next time you have a blood test, ask them to take it all out."

In any survey research, the quality of the sample has a tremendous impact on the accuracy of the information it provides. All samples can be divided into one of two classes: probability, and nonprobability samples. They differ in how they identify who will actually be included in the sample. *Probability* samples, sometimes called *random samples,* use a process of random selection that allows every member of the population to have an equal, or known, probability of selection. Although probability samples are more expensive and time consuming to construct, researchers generally have more confidence in them. *Nonprobability* samples, in which membership is determined by happenstance or convenience, are more likely to produce biased results.

All the ratings companies described in this book are trying to achieve, or at least to approximate, the virtues of probability sampling. Their technical documents are laced with the language of probability samples. To acquire the needed working vocabulary, therefore, one must be familiar with the principles of probability sampling. The following discussion is designed to provide that familiarity, in a way that does not assume a background in quantitative methods on the part of the reader. Those already familiar with sampling may wish to skip to the section on measurement.

Basic Concepts

Sampling begins with a definition of the population being studied. This requires a decision about what kind of things are to be studied—called elements or units of analysis in the parlance of researchers—and which of those things constitute the relevant population. In ratings research, units of analysis are either people or households. Because the use of radio is thought to be a rather individualistic, one-on-one experience, radio ratings have long used people as the unit of analysis. Television ratings have historically used households as the unit of analysis, although that is changing with the introduction of the new measurement technologies discussed later.

Whatever the unit of analysis, they can be grouped to create a larger *population* or *universe*. Researchers must define these populations so they can tell who belongs in a given population. For example, if we were attempting to create national television ratings, all households in the United States with one or more sets might be appropriate. Local markets are often more of a problem, since we must determine who lives, say, in Washington, DC, as opposed to Baltimore. As a practical matter, the ratings services create markets (called either DMAs or ADIs) by using counties as building blocks. They do this by determining which stations the people in a particular county listen to, and assigning counties accordingly.

The next step is to obtain a complete list of all the elements included in the population. That list is called a *sampling frame*. It is from the sampling frame that specific elements will be identified for inclusion in the sample. For example, if we have a list of all the television households in Baltimore (assume 1 million for convenience), and randomly picked one home, we would know that it had a one-in-a-million chance of selection, just like every other home in the population. Hence, we would have met the basic requirement of probability sampling. All we would have to do, then, is repeat the process until we had a sample of the desired size.

The procedure we have just described produces a *simple random sample*. Despite its conceptual elegance, this sort of sampling technique is not often used in ratings research. One reason is that the real world is less cooperative than this approach to sampling assumes. In addition, there are more efficient and powerful sampling designs available to researchers. The most common sampling techniques of the ratings companies are described here.

Sample Designs

Systematic Random Sampling. One probability sampling technique that involves only a minor variation on simple random sampling is called *systematic random sampling*. Like a simple random sample, this approach requires the use of a sampling frame. Usually, ratings firms will buy sampling frames from companies whose business it is to maintain and sell such lists. Metromail is the company that Arbitron has used for many years. Nielsen does its own. Typically, these frames are lists of telephone households. Homes with unlisted numbers can be included through the use of randomly generated numbers. Frames that have been amended in this way are called *expanded* or *total* sampling frames.

Once an appropriate frame is available, systematic sampling is straight forward. Because you have a list of the entire population, you know how large it is. You also know how large a sample you want from that population. Dividing population size by sample size lets you know how often you have to pull out a name or number as you go down the list. For example, suppose you had a population of 10,000 individuals and you wanted to have a sample of 1,000. If you started at the beginning of the list and selected very 10th name, you would end up with a sample of the desired size. That "nth" interval is called the *sampling interval*. The only further stipulation for systematic sampling, and it is an important one, is that you pick your starting point at random. In that way, everyone has had an equal chance of being selected, again meeting the requirement imposed by probability sampling.

Systematic sampling, as it is practiced by the ratings companies, is not perfect. For one thing, an absolutely complete list of the population is almost impossible to obtain. People living in temporary or group housing may be hard to track down. In many markets, a substantial portion of households are without a telephone. If lists are limited to homes with telephones, some people will be under-represented in the final ratings report. Conversely, households with more than one telephone number may have a greater probability of selection than other homes. Any of these factors can introduce biases into samples.

Multi-Stage Cluster Sampling. Fortunately, not all probability samples require a complete list of every single element in the population. One sampling procedure that avoids that problem is called *multi-stage cluster sampling*. Cluster sampling repeats two processes: listing and sampling. Each two-step cycle constitutes a stage. Although systematic random sampling is a one-stage process, multi-stage cluster sampling, as the name implies, goes through several stages.

A ratings company might well use multi-stage sampling to identify a national sample. After all, coming up with a list of every single household in the nation would be quite a chore. However, it would be possible to list every single county in the United States. If that were done, the research company could then draw a random sample of counties. In fact, this is essentially what Nielsen does to begin the process of creating a national sample of U.S. households. After that, census tracts within those selected counties could be listed and randomly sampled. Third, city blocks within selected census tracts could be listed and randomly sampled. Finally, with a manageable number of city blocks identified, researchers might be placed in the field, with specific instructions, to find individual households for participation in the sample.

Because, in this example, the clusters that are listed and sampled at each stage are geographic areas, this type of sampling is sometimes called a *multi-stage area probability sample*. Despite the laborious nature of such sampling techniques, compared to the alternatives, they offer important advantages. Specifically, no sampling frame listing every household is required, and researchers in the field can contact households even if they do not have a telephone.

However, a multi-stage sample is more likely to be biased than a single-stage sample. This is because, through each round of sampling, a certain amount of error accompanies the selection process. The more the stages, the more the possibility of error. For example, suppose that during the sampling of counties described earlier, areas from the Northwestern United States were over-represented. That could happen just by chance, and it would be a problem carried through subsequent stages. Now suppose that bias is compounded in the next stage by the selection of Census tracks from a disproportionate number of affluent areas. Again, that is within the realm of chance. Even if random selection is strictly observed, a certain amount of "sampling error" creeps in. We discuss this more fully later in this chapter when we cover sources of error.

Stratified Sampling. Some sorts of error can be minimized by using a third kind of sampling procedure called *stratified sampling*. This is one of the most powerful sampling techniques available to survey researchers. Stratified sampling requires the researcher to group the population being studied into relatively homogeneous subsets, called *strata*. Suppose we have a sampling frame that indicated the gender of everyone in the population. We could then group the population into males and females, and randomly sample the appropriate number from each strata. By combining these subsamples into one large group, we would have created a probability sample that has exactly the right proportions of men and women. Without stratification, that factor would have been left to chance. Hence, we have improved the representativeness of the sample. That added precision could be important if we were studying things related to gender, like watching sports on TV, or certain product purchases like cosmetics and tires.

Stratified sampling obviously requires that the researcher have some relevant information about the elements in a sampling frame (e.g., the gender of everyone in the population). In single-stage sampling that is sometimes not possible. In multi-stage sampling, there is often an abundance of information because we tend to know more about the large clusters we begin with. Consider, again, the process that began by sampling counties. Not only could we list all U.S. counties, but we could group them by the state or region of the country they are in, the size of their populations, and so forth. Other sorts of groupings could be used at subsequent stages in the process. By combining stratification with multi-stage cluster sampling, therefore, we could increase the representativeness of the final sample. That is just what most ratings services will do.

Cross-Sectional Surveys. All of the sample design issues we have discussed thus far have dealt with how the elements in the sample are identified. Another aspect of sample design deals with how long the researcher actually studies the population or sample. *Cross-sectional* surveys occur at a single point in time. In effect, these studies take a "snapshot" of the population. Much of what is reported in a single ratings book, could be labeled cross-sectional. Such studies may use any of the sampling techniques just described. They are alike insofar as they tell you what the population looks like now, but not how it has changed over time. Information about those changes can be quite important. For instance, suppose the ratings book indicates that your station has an average rating of 10. Is that cause for celebration or dismay? The answer depends on whether that represents an increase or decrease in the size of your audience, and true cross-sectional studies will not tell you that.

Longitudinal Studies. These studies are designed to provide you with information about changes over time. In ratings research, there are two kinds of longitudinal designs in common use; trend studies and panel studies. A *trend study* is one in which a series of cross-sectional surveys, based on independent samples, is conducted on a population over some period of time. The definition of the population remains the same throughout the study, but individuals may move in and out of the population. In the context of ratings research, trend studies can be created simply by considering a number of market reports done in succession. For example, tracing a station's performance across a year's worth of ratings books constitutes a trend study. People may have moved to or from the market in that time, but the definition of the market (i.e., the counties assigned to it) has not changed. Most market reports, in fact, provide some trend information from past reports. *Panel studies* draw a single sample from a population, and continue to study that sample over time. The best example of a panel study in ratings research involves the metering of people's homes. This way of gathering ratings information, which we describe later in the chapter, may keep a household in the sample for years.

Sources of Error

One of the principle concerns that both the users and producers of ratings have is with error in the data. The concept of "error" is not just a matter of mistakes being made, more broadly, it addresses the extent to which ratings information fails to report what is actually happening in the population. Error is the difference between what the ratings estimate to be true, and what is true. An understanding of where error comes from, and how the ratings companies do or do not deal with it, is one of the characteristics of a sophisticated ratings user.

There are four sources of error in ratings data: sampling error, nonresponse error, response error, and processing error. The first two involve sampling, and so, are dealt with here. The last two involve measurement and the production process, respectively, and are covered in the sections that follow.

Sampling Error. This is the most abstract of the different kinds of error we discuss. It is a statistical concept that is common to all survey research. Basically, it involves a recognition that as long as we try to estimate what is true for a population by studying something less than the entire population, there is a chance that we will miss the mark. Even if we use very large, perfectly executed random samples, it is possible that they will fail to accurately represent the populations from which they were drawn. This is inherent in the process of sampling. Fortunately, if we employ random samples, we can, at least, use the laws of probability to make statements about the amount of sampling error we are likely to encounter.

The best way to explain sampling error, and a host of terms that accompany the concept, is to work our way through a hypothetical study. Suppose that the Super Bowl was played yesterday and we wanted to estimate what percent of households actually watched the game (i.e., the game's rating). Let us also suppose that the "truth" of the matter is that exactly 50% of U.S. homes watched the game. Of course, ordinarily we would not know that, but we need to assume this knowledge to make our point. The true population value is represented in Fig. 5.1A.

In order to estimate the game's rating, we decide to draw a random sample of 100 households from a list of all the television households in the country. Because we have a complete sampling frame (how convenient!), every home has had an equal chance to be selected. Next, we call each home and ask if they watched the game. Because they all have telephones, perfect memories, and are completely truthful (again, convenient), we can assume we have accurately recorded what happened in these sample homes. After a few quick calculations, we discover that only

RATINGS RESEARCH METHODS

A

[Scale from 0 to 100 in increments of 10, with arrow pointing to 50 labeled "True Population Value"]

The scale includes all the possible ratings for the Super Bowl, and indicates the true population value is 50.

B

[Scale from 0 to 100 with an "X" marked above 46, labeled "Sample 1 = 46"]

The first random sample drawn from the population indicated that 46% watched the game. This estimate is marked with an "X."

C

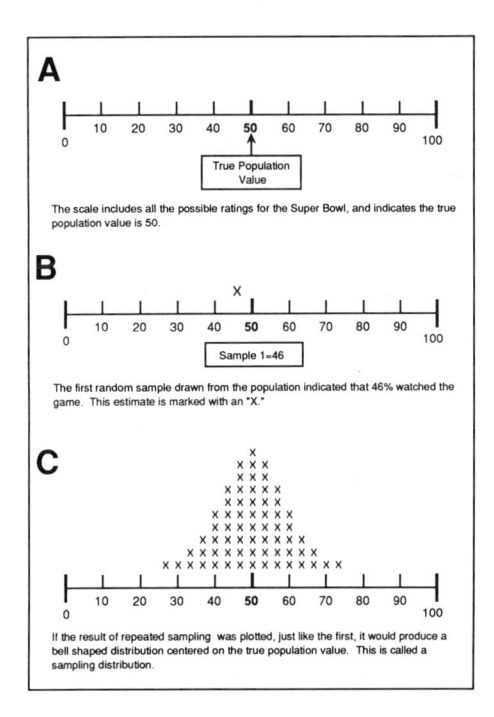

[Bell-shaped distribution of X's centered on 50, plotted above scale from 0 to 100]

If the result of repeated sampling was plotted, just like the first, it would produce a bell shaped distribution centered on the true population value. This is called a sampling distribution.

FIG. 5.1. A sampling distribution.

46% of those we interviewed saw the game. This result is plotted in of Fig. 5.1B.

Clearly, we have a problem here. Our single best guess as to how many homes saw the game is 4% lower than what was, in fact, true. In the world of media buying, 4 ratings points can mean a lot of money. It should, nevertheless, be intuitively obvious that even with our convenient assumptions and strict adherence to sampling procedures, such a disparity is entirely possible. In fact, it would have been surprising to hit the nail on the head the first time out. That 4% difference we have observed does not mean we did anything wrong, it is just a sampling error.

Because we have the luxury of a hypothetical case here, let's assume that we repeat the sampling process. On our second time out, 52% of the sample say they watched the game. Better, but still in error, and still a plausible kind of occurrence. Finally, suppose that we draw 1,000 samples, just like the first two. Each time we plot the result of that sample. If we did this, the result would look something like Fig. 5.1C.

The shape of this figure reveals a lot, and is worth considering for a moment. It is a special kind of frequency distribution that a statistician calls a *sampling distribution*. In our case it forms a symmetrical, bell-shaped curve indicating that, when all was said and done, more of our sample estimates hit the true population value (i.e., 50%) than any other single value. It also indicates that although most of the sample estimates clustered close to 50%, a few were way off. In essence, this means that, if you use probability sampling, reality has a way of anchoring your estimates and keeping most of them fairly close to what is true. It also means that sooner or later, you are bound to hit a clunker.

What is equally important about this sampling distribution is that it will assume a known size and shape. The most frequently used measure of that size and shape is called the *standard error* (SE). For those familiar with introductory statistics, this is very much like a "standard deviation." It is best conceptualized as a unit along the baseline of the distribution. Fig. 5.2 gives the simplest formula for calculating standard error with ratings data.

What is remarkable about the standard error and what you will have to accept on faith unless you want to delve much more deeply into calculus, is that when it is laid out against its parent sampling distribution, it will bracket a precise number of samples. Specifically, plus or minus one SE will always encompass 68% of the samples in the distribution. Plus or minus 2 SE (technically, that should be 1.96), encompasses 95% of all samples. In our example, the SE works out to be approximately 5 ratings

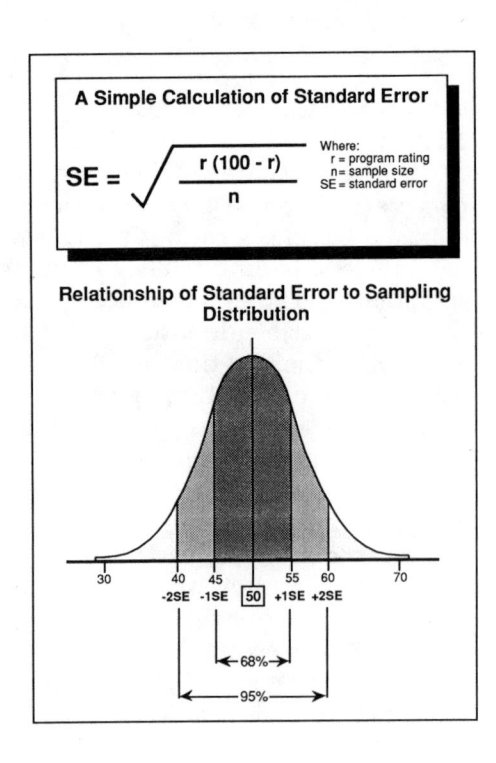

A Simple Calculation of Standard Error

$$SE = \sqrt{\frac{r(100-r)}{n}}$$

Where:
r = program rating
n = sample size
SE = standard error

Relationship of Standard Error to Sampling Distribution

FIG. 5.2. Standard error.

points, which means that 68% of the hypothetical samplings will have produced results between 45% and 55% (i.e., 50% plus or minus 5%). That relationship between SE and the sampling distribution is depicted in Fig. 5.2.

None of this would be of interest to anyone other than a mathematician, were it not for the fact that such reasoning provides us with a way to make statements out the accuracy of ratings data. Remember that in our first sample, we found 46% watching the Super Bowl. Ordinarily, that would be our single best guess about what was true for the population. We would recognize, however, that there was a possibility of sampling error, and we would want to know the odds of the true population value being something different than our estimate. We could state those odds by using our estimated rating (i.e., 46) to calculate SE, and placing a bracket around our estimate, just like the one in Fig. 5.2. The resulting statement would sound like this, "We estimate that the Super Bowl had a rating of 46, and we are 95% confident that the true rating falls between 36 and 56."

The range of values given in that statement (i.e., 36 to 56) is called the *confidence interval*. Confidence intervals are often set at plus or minus two SE, and will therefore have a high probability of encompassing the true population value. When you hear someone qualify the results of a survey by saying something like, "these results are subject to a sampling error of plus or minus 3%," they are giving you a confidence interval. What is equally important, but less often heard, is how much confidence should be placed in that range of values. To say we are "95% confident," is to express a *confidence level*. At the 95% level, we know that 95 times out of 100 the range we report will include the population value. Of course, that means that 5% of the time we will be wrong, because it is always possible our sample was one of those clunkers. But at least we can state the odds, and satisfy ourselves that an erroneous estimate is a remote possibility.

Such esoteric concepts take on practical significance because they go to the heart of the ratings accuracy. For example, reporting that a program has a rating of 15, plus or minus 10, leaves a lot of room for error. Even fairly small margins of error (e.g., SE = 1), can be important if the estimates they surround are themselves small (e.g., a rating of 3). That is one reason why ratings services will routinely report *relative standard error* (i.e., SE as a percentage of the estimate) rather that the absolute level of error. In any event, it becomes critically important to reduce sampling error to an acceptable level. Three factors affect the size of that error. One is beyond the control of researchers, two are not.

The source of sampling error that we cannot control has to do with the population itself. Some populations are just more complicated that others.

A researcher refers to these complexities as "variability" or "heterogeneity" in the population. To take an extreme case, if everyone in the population were exactly alike (i.e., perfect homogeneity), then a sample of one person would suffice. Unfortunately, media audiences are not homogeneous, and to make matters worse, they are getting more heterogeneous all the time. Think about how television has changed over the years. It used to be that people could watch the three networks, maybe an independent or public station, and that was it. Now most homes have cable or VCRs, as well as more stations to choose from. All other things being equal, that makes it more difficult to estimate who is watching what.

The two factors that ratings companies can use to reduce sampling error involve the sample itself. *Sample size* is the most obvious, and important of these. Larger samples reduce the magnitude of sampling error. Its just common sense that we should have more confidence in results from a sample of 1,000, than 100. What is counterintuitive is that sample size and error do not have a one-to-one relationship. That means doubling the size of the sample does not cut the SE in half. Instead, you must quadruple sample to reduce the SE by half. You can satisfy yourself of this by looking back at the calculation of SE in Fig. 5.2. To reduce the SE from 5 to 2.5, you must increase the sample size from 100 to 400. You should also note that the size of the population you are studying has no direct impact on the error calculations. All other things being equal, small populations require samples just a big as large populations.

These aspects of sampling theory are more than just curiosities. They have a substantial impact on the conduct and economics of the ratings business. For example, although it is always possible to improve the accuracy of the ratings by increasing the size of the samples on which they are based, you very quickly reach a point of diminishing returns. This was nicely demonstrated in research conducted by CONTAM, an industry group formed in response to the Congressional hearing of the 1960s. That study collected viewing records from over 50,000 households around the country. From that pool, 8 sets of 100 samples were drawn. Samples in the first set had 25 households each. Sample sizes for the following sets were: 50, 100, 250, 500, 1,000, 1,500, and 2,500. The results are shown in Fig. 5.3.

At the smallest sample sizes, individual estimates of the "Flintstones" audience varied widely around the actual rating of 26. Increasing sample sizes from these low levels produced dramatic improvements in the consistency and accuracy of sample estimates, as evidenced in tighter clustering. For example, going from 100 to 1,000 markedly reduced sampling error and only required adding 900 households. Conversely, going from 1,000 to 2,500 resulted in a modest improvement, yet it required an increase of 1,500 households. Such relationships mean ratings companies and their

THE FLINTSTONES
U.S. Total Population
March 1963

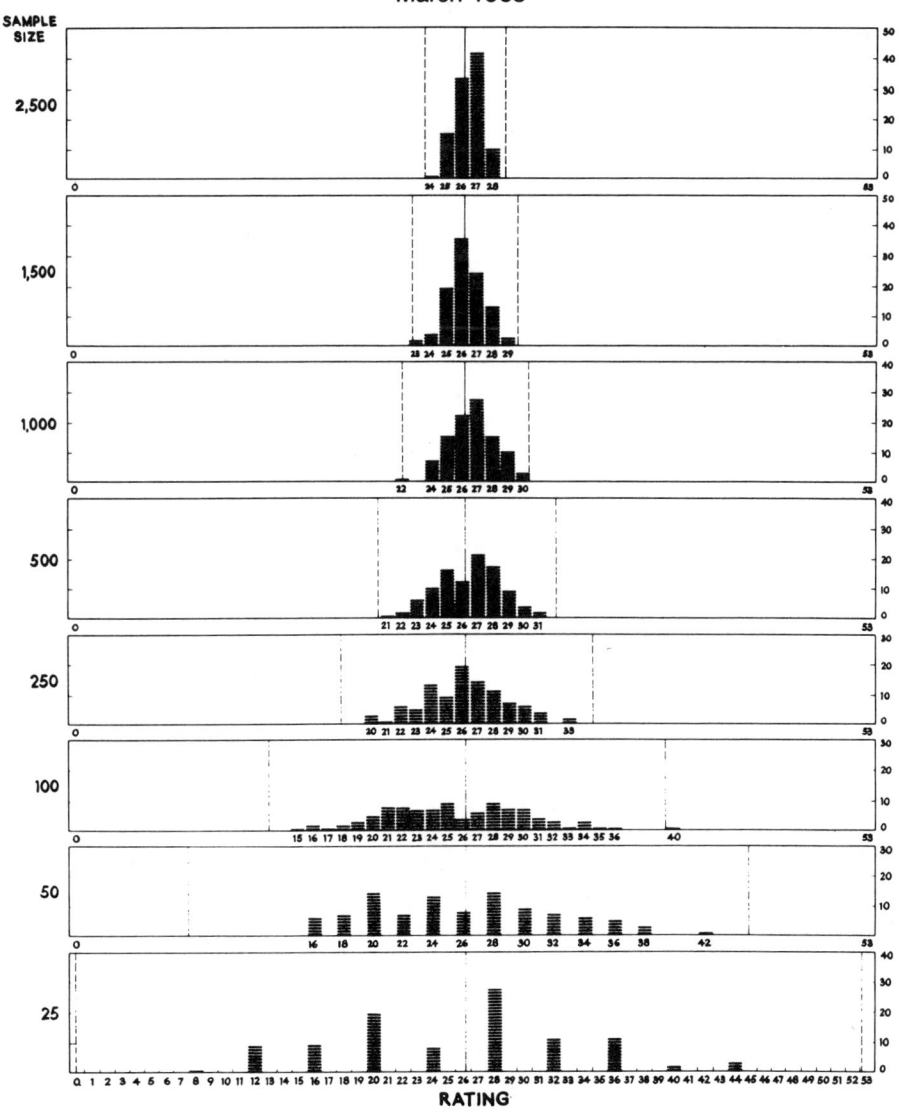

| True population rating
: Point outside of which 2.6 out of 1,000 sample results should fall according to theory (3σ).
— Represents one sample yielding indicated rating estimate.
Distribution of Sample Ratings Results Based on 100 Diary Samples Each of 8 Different Sizes: 25 to 2500.
Source: CONTAM Study, No. 1.

FIG. 5.3. Effect of sample size on sampling error (from Beville, 1988).

clients have to strike a balance between the cost and accuracy of ratings data.

In practice, several other things determine the sample sizes that a ratings company will use. As suggested earlier, more complex populations will require larger samples to achieve a certain level of sampling error. This has meant that radio requires bigger samples than television, because there have been more radio stations to fragment the audience. Similarly, if you intend to study relatively small segments of the audience (e.g., mean 18 to 21) you will require larger overall samples. And even though larger populations do not, theoretically, need bigger samples, because of their relative complexity and the volume of media dollars available, larger markets are studied with larger samples.

The only other factor that the ratings company can employ to reduce sampling error is to improve the design of the sample. For reasons that we have already discussed, certain kinds of probability samples, like stratified samples, are more accurate than others. This strategy is commonly used, but there is a limit to what can be achieved. We should also note that when these more complex sample designs are used, the calculation of SE becomes a bit more involved than Fig. 5.2 indicates. We address those revised computations later.

Nonresponse Error. This is the second major source of error we encounter in the context of sampling. It occurs because not everyone we might wish to study will cooperate or respond. Remember, our entire discussion of sampling error assumed that everyone we wanted to include in the sample gave us the information we desired. In the real world that just does not happen. To the extent that those who do not respond are different from those who do, there is a possibility that the samples we actually have to work with may be biased. Many of the procedures that the ratings services use represent attempts to correct nonresponse error.

The magnitude of nonresponse error varies from one ratings report to the next. The best way to get a sense of it is to look at the response rate the ratings service reports. Every ratings company will identify an original sample of people or households that it wishes to use in the preparation of its ratings estimates. This ideal sample is usually called the *initially designated sample*. Some members of the designated sample, however, will refuse to cooperate; others will agree to be in the sample, but then fail to provide complete information. In other words, many will not respond as hoped. Obviously, only those who do respond can be used to tabulate the data. The latter group constitutes what is called the *in-tab sample*. The *response rate* is simply the percent of people from the initially designated sample who actually gave the ratings company useful information.

Different techniques for gathering ratings data are associated with different response rates. Telephone surveys, for example, tend to have relatively high response rates. The most common measurement techniques, like placing diaries or meters, will often produce response rates in the neighborhood of 50%. Furthermore, different measurement techniques work better with some kinds of people than others. The nonresponse errors associated with measurement are discussed in the next section.

Because nonresponse error has the potential to bias the ratings, research companies employ one of two general strategies to minimize or control it. First, you can take action before the fact to improve the representativeness of the in-tab sample. Second, you can make adjustments in the sample after data have been collected. Usually both strategies are employed. Either way, you need to know what the population looks like in order to judge the representativeness of your in-tab sample and to gauge the adjustments that are to be made.

Population or *universe estimates,* therefore, are essential in correcting for nonresponse error. Determining what the population looks like (i.e., age and gender breakdowns, etc.) is usually based on U.S. Census information. The Census is updated only every 10 years—but parts are revised, based on sampling, more frequently. Ratings companies often buy more current universe estimates from other research companies. Market Statistics, Inc., is one such company that supplies both Arbitron and Nielsen. Occasionally, certain attributes of the population that have not been measured by the Census Bureau, like cable penetration, must be estimated. To do this, it may be necessary to conduct a special *enumeration study* that establishes important universe estimates.

Once it is known what targets to shoot for, corrections for nonresponse error can be made. Before-the-fact remedies include the use of special recruitment techniques and buffer samples. The most desirable solution is to get as many of those in the originally designated sample as possible to cooperate. Doing so requires a deeper understanding of the reasons for nonresponse, and combating those with counteractive measures. For example, ratings services will often provide sample members with some monetary incentive. Perhaps different types of incentives will work better or worse with different types of people. Following up on initial contacts or making sure that interviewers and research materials are in a respondent's primary language will also improve response rates. The major ratings companies are aware of these alternatives, and on the basis of experience, know where they are likely to encounter nonresponse problems. Arbitron, for example, will use what it calls a *differential survey treatment* for Black and Hispanic households. These are special recruitment techniques they use to improve minority representation in the sample.

If improved recruitment fails to work, under-represented groups can be increased by additional sampling. *Buffer samples* are simply lists of additional households that have been randomly generated and are held in reserve. If, as sampling progresses, it becomes apparent that responses in one county are lagging behind expectations, the appropriate buffer sample can be enlisted to increase the size of the sample drawn from that area. A similar procedure might be used by field workers if they encounter a noncooperating household. In such an event, they would probably have instructions to sample a second household in the same neighborhood, perhaps even matching the noncooperator on key household attributes.

Once the data are collected, another technique can be used to adjust for nonresponders. *Sample weighting,* sometimes called *sample balancing,* is a statistical procedure that gives the responses of certain kinds of people more influence over the ratings estimates than their numbers in the sample would suggest. Basically, the ratings companies compare the in-tab sample and the universe estimates (usually on geographic, ethnic, age and gender breakdowns), and determine where they have too many of one kind of person, and not enough of another. Suppose, for example, that 18- to 24-year-old men accounted for 6% of the population, but only 3% of the in-tab sample. One remedy for this would be to let the responses of each young man in the in-tab count twice. Conversely, the responses of over-represented groups would count less than once. The way to determine the appropriate weight for any particular group is to divide their proportion in the population by their proportion in the sample (e.g., 6% / 3% = 2).

If you think the use of buffer samples or weighting samples is not a completely adequate solution to problems of nonresponse, you are right. Although these procedures may make in-tab samples look like the universe, they do not eliminate nonresponse error. The people who are drawn through buffer samples or whose responses count more than once might still be systematically different from those who did not cooperate. That is why some people question the use of these techniques. The problem is that failing to make these adjustments also distorts results. For example, if you programmed a radio station that catered to 18- to 24-year-old men, you would be unhappy that they tend to be under-represented in most in-tab samples, and probably welcome the kind of weighting just described above, flaws and all. Today, the accepted industry practice is to weight samples. We return to this topic when we discuss the process of producing the ratings.

The existence of nonresponse error, and certain techniques used to correct for such error, means that the samples the ratings services actually use are not perfect probability samples. That fact, in combination with the use of relatively complex sample designs, means that calculations of standard error are a bit more involved than our earlier discussion indicated. Without going into detail, error is affected by the weights in the sample, whether you are dealing with households or persons, and whether you are estimating the audience at a single point in time or the average audience over a number of time periods. Further, actual in-tab sample sizes are not used in calculating error. Rather, the ratings services derive what they call *effective sample sizes* for purposes of calculating SE. These take into account the fact that their samples are not simple random samples. Effective sample sizes may be smaller than, equal to, or larger than actual sample sizes. No matter the method for calculating SE, however, the use and interpretation of that number is as described earlier.

MEASUREMENT

Although sampling is an essential aspect of the ratings business, so too is measurement. It is one thing to identify who you want to study, it is quite another to record what they see on television or what they hear on the radio. The latter activity is referred to as *measurement*.

Technically, *measurement* is defined as a process of assigning numbers to objects, according to some rule of assignment. The "objects" that the ratings companies are usually measuring are people, although as we have seen, households can also be the unit of analysis. The "numbers" simply quantify the characteristics or behaviors that we wish to study. This kind of quantification makes it easier to manage the relevant information and to summarize the various attributes of the sample. For example, if a person saw the "CBS Evening News" last night, we might assign him or her a "1." Those who did not see the news might be assigned a "0." By reporting the percentage of 1s we have, we could produce a rating for the CBS news. The numbering scheme that the ratings services actually use is a bit more complicated than that, but in essence, that is what goes on.

Researchers who specialize in measurement are very much concerned with the accuracy of the numbering scheme they use. After all, anyone can assign numbers to things, it is making sure that the numbers capture something meaningful that is the real trick. Researchers express their concerns about the accuracy of a measurement technique with two concepts: reliability and validity. *Reliability* is the extent to which a measurement procedure will produce consistent results in repeated applications. In other words, if what you are trying to measure does not change, then an accurate measuring device should end up assigning it the same number time after time. If that is the case, the measure is said to be reliable. Just because a measurement procedure is reliable, however, does not mean that it is completely accurate, it must also be valid. *Validity* is the extent to which a measure actually quantifies the characteristic it is supposed to quantify. For instance, if we wanted to measure a person's program preferences, we might try to do so by recording which shows he or she watches most frequently. This approach might produce a very consistent, or reliable, pattern of results. However, it does not necessarily follow that the program a person sees most often is their favorite. Scheduling, rather than preference, might produce such results. Therefore, measuring preferences by using a person's program choices might be reliable, but not particularly valid.

Definitional Issues

One of the first questions that must be addressed in any assessment of measurement techniques is, "What are you trying to measure?" Confusion on this point has led to a good many misunderstandings about ratings data. At first glance the answer seems simple enough. Ratings measure exposure to the electronic media. But even that definition leaves much unsaid. To think this through, two factors need to be more fully considered: (a) What do we mean by "media"? and (b) What constitutes exposure?

Defining the media side of the equation raises a number of possibilities. It might be, for example, that we have no interest in the audience for specific content. As we noted in the previous chapter, some effects researchers are only concerned with how much television people watch overall. Although knowing the amount of exposure to a medium might be useful in some applications, it is not terribly useful to advertisers. Exposure to a certain channel or station is another possibility that is not content specific. Radio station audiences and, to a certain extent, cable network audiences, are reported this way. Here the medium may be no more precisely defined than attendance during a broad daypart, or an average quarter hour.

In television ratings, exposure is usually tied to a specific program. Here, too, however, definitional questions can be raised. How much of a program must a person see before they are to be included in that program's audience? If a few minutes is enough, then the total audience for the show will probably be larger than the audience at any one point in time. Some of the measurement techniques we discuss in the following section are too insensitive to make such minute-to-minute determinations, but for other approaches, this is far from being a moot point.

Advertisers are, of course, most interested in who sees their commercials. So, a case can be made that the most relevant way to define the media is not program content, but commercial content. Such "commercial ratings" are not routinely produced by the major ratings services, but newer measurement technologies raise the possibility that the audience for brief commercial messages could be quantified.

The other aspect of this definitional question is determining what is meant by exposure. Once again, there are a number of possibilities. *Exposure* is usually defined as the choice of a particular station or program. Under this definition, the only thing that is relevant is who is present when the set is in use. In fact, some measurement techniques are incapable of recording who is in the room. Once it has been determined that audience members have tuned to a particular station, further questions about the quality of exposure are left unanswered.

It is well documented, however, that much of our media use is accompanied by other activities. People may read, talk, eat, play games, or do the dishes while the set is in use. Whatever the case, it is clear that during a large portion of the time that people are in the audience, they are not paying much attention. This has lead many commentators to argue that defining exposure as a matter of choice greatly overstates people's real exposure to the media. An alternative, of course, would be to stipulate that exposure must mean that a person is paying attention to the media, or perhaps even understanding what is seen or heard. Despite the logic of this definition, measuring a person's level of attention or perception is extremely difficult to do in an efficient, valid way.

Another shortcoming that critics of the ratings services have raised from time to time is that operational definitions of exposure tell us nothing about the quality of the experience in a more affective sense. For example, do people like what they see, or find it to be informative and enlightening? *Qualitative ratings* such as these have been produced on an irregular basis, not so much as a substitute for existing services, but rather as a supplement. In the early 1980s, the Corporation for Public Broadcasting, in collaboration with Arbitron, conducted field tests of such a system. More recently, an independent Boston-based company named Television Audience Assessment tried selling qualitative ratings information. That effort failed, and at present, there does not seem to be enough demand for qualitative ratings to sustain their continuous production.

Obviously, these definitional questions help determine what the ratings really are and how they are to be interpreted. If different ratings companies used vastly different definitions of exposure to media, their cost structures and research products might be quite different as well. The significance of these issues has not been lost on the affected industries. In 1954, the Advertising Research Foundation (ARF), released a set of recommendations that took up many of these issues. In addition to advocating the use of probability samples, ARF recommended that tuning behavior be the accepted definition of exposure. That standard has been the most widely accepted, and has effectively guided the development of the measurement techniques we use today.

Measurement Techniques

There are several techniques that the ratings services use to measure people's exposure to electronic media. Each has certain advantages and disadvantages. The biases of each technique contribute to the third kind of error we mentioned earlier (i.e., response error). *Response error* includes inaccuracies contained in the responses generated by the measurement procedure. We discuss, in general terms, each major approach to audience measurement. So as not to get too bogged down in details, we may gloss over differences in how each ratings company operationalizes a particular scheme of measurement. Here again, the reader wishing more information should see each company's description of methodology.

Diaries are the most widely used of all measurement techniques. Although they are no longer employed to estimate national network audiences, huge numbers of diaries are used to determine local radio and television audiences. In one television ratings sweep alone, Arbitron will collect diaries from over 100,000 households. Nielsen, too, uses diaries, and will gather another 100,000 to produce television ratings reports in most markets around the country.

A diary is a small paper booklet in which the diary keeper is supposed to record his or her media use for a 1-week period. To produce television ratings, one diary is kept for each TV set in the household. Figure 5.4 is the first page from an Arbitron television diary. It begins on Wednesday at 6 a.m. and thereafter divides the day into quarter-hour segments ending a 2 a.m. Each of the remaining days of the week is similarly divided. During each quarter hour that the set is in use, the diary keeper is supposed to note the relevant call letters, channel number, and program title, as well as which family members and/or visitors are watching. The diary concludes with a few additional questions about household composition, and the channels that are received in the home.

Diaries are also used to measure radio audiences. Radio diaries, however, are supposed to accompany people rather than sets. That way, an individual can record listening that occurs outside the home. Arbitron is the only ratings research firm that uses diaries to produce radio audience estimates. Figure 5.5 is the first page of an Arbitron radio diary. It begins on Thursday, and divides the day into broader dayparts than the rigid quarter-hour increments of the TV diary. Because a radio diary is a personal record, the diary keeper does not note whether other people were listening. The location of listening, however, is recorded.

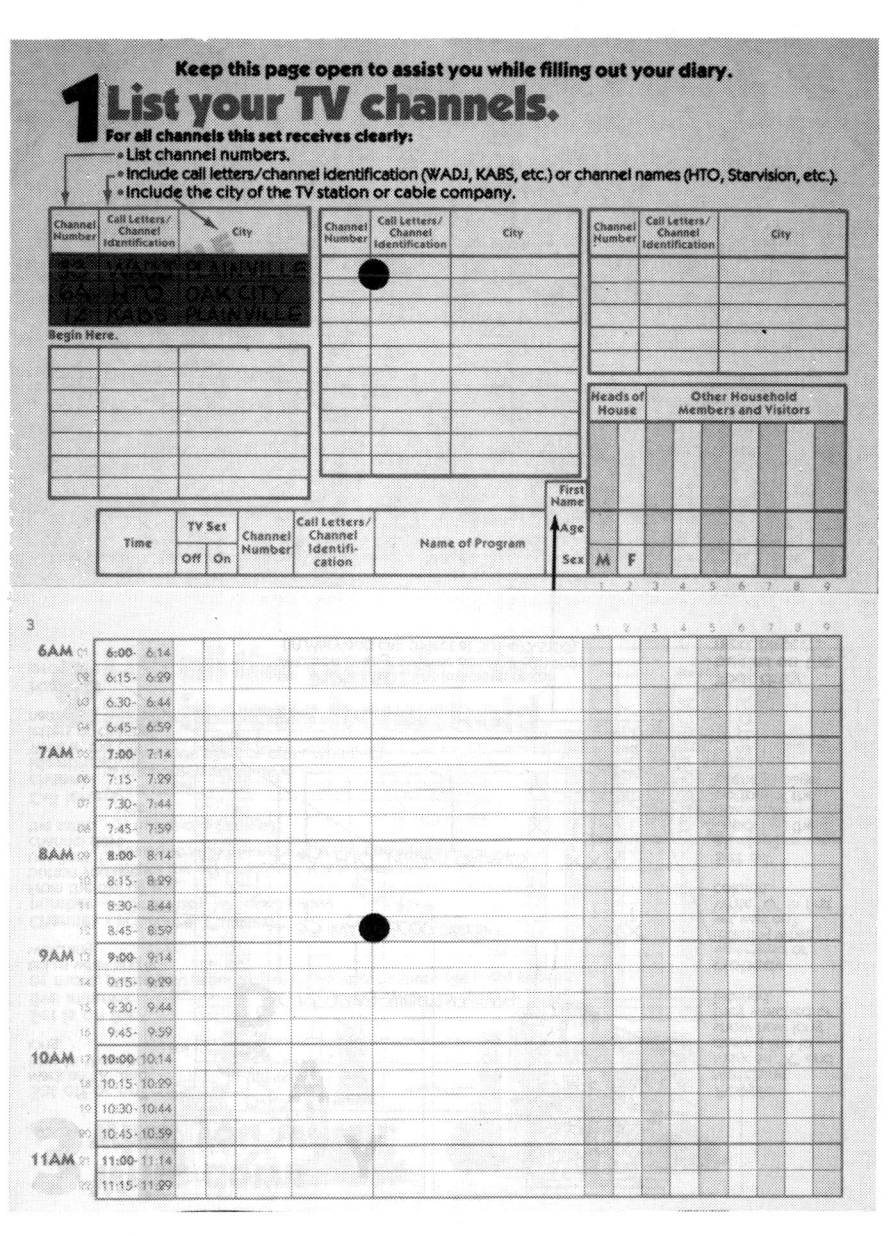

FIG. 5.4. Arbitron television diary (reprinted with permission of the Arbitron Company).

103

THURSDAY

	Time		Station			Place			
			Call letters or station name	Check (✓) one		Check (✓) one			
	Start	Stop	Don't know? Use program name or dial setting.	AM	FM	At Home	In a Car	At Work	Other Place
Early Morning (from 5 AM)									
Midday									
Late Afternoon									
Night (to 5 AM Friday)									

If you didn't hear a radio today, please check here. ☐

FIG. 5.5. Arbitron radio diary (reprinted with permission of the Arbitron Company).

Diary placement and retrieval techniques vary, but the usual practice goes something like this. Members of the initially designated sample are called on the telephone by the ratings company so that it can secure the respondent's cooperation, and collect some initial information. Those who are to be excluded (e.g., people living in group quarters), or those who will receive special treatment (e.g., Spanish-speaking households) are identified at this stage. Follow-up letters may be sent to households that have agreed to cooperate. Diaries are, then, either mailed or delivered to the home in person by field personnel. Incidentally, although respondents are asked to cooperate, diaries are distributed to all who are contacted even if they say they are not interested in cooperating. And, the response rate is just about as high for those who initially say they are unwilling as for those who agree. Quite often, a monetary incentive of $1 or so is provided as a gesture of "goodwill," although goodwill is more likely to be used in certain markets that have traditionally had lower response rates. During the week, another letter or phone call may encourage the diary keeper to note his or her media use. Diaries are designed to be sealed and placed directly in the mail, which is typically how the diary is returned to the ratings company at the end of the week. Occasionally, a second monetary reward follows the return of the diary. In some special cases, homes are called and the diary information is collected over the telephone.

Diaries have some significant advantages that account for their popularity. They offer a relatively inexpensive method of data collection. Considering the wealth of information that a properly filled out diary contains, none of the techniques we discuss here is as cost effective. Most importantly, they report which people were actually in the audience. In fact, until recently, diaries had to be used in conjunction with more expensive metering techniques to determine the demographic composition of the television audience. Even if the newer peoplemeters become the standard in large media markets, it seems likely that diaries will continue to be used in most local markets.

Nevertheless, there are a number of disadvantages associated with the use of diaries, problems of both nonresponse and response error. We have already discussed nonresponse error in the context of sampling. It should be noted, however, that diaries are particularly troublesome in this regard. Response rates on the order of 50% are common, and in some markets will drop below that. Obviously, diary keepers must be literate, but methodological research undertaken by the industry suggests that those who fill out and return diaries are systematically different in other ways. Younger people, especially younger males, are less responsive to the diary technique. Blacks, too, are less likely to complete and return a diary. There is also some evidence that those who return a television diary are heavier users of the medium than nonrespondents.

There are a number of response errors typical of diary data as well. There is a fair amount of anecdotal evidence that diarykeepers frequently do not note their media use as it occurs, but try to recollect it at the end of the day or the week. To the extent that entries are delayed, errors of memory are more likely. Similarly, it appears that diary keepers are more diligent in the first few days of diary keeping than the last. This "diary fatigue" may artificially depress viewing or listening levels on Mondays and Tuesdays. Viewing late at night, viewing of short duration, viewing of less well-known programming, and viewing of secondary sets (e.g., in bedrooms, etc.) is typically under-reported. Children's use of the television, at times when an adult diary keeper is not present, is also likely to go unreported.

These are significant, if fairly benign, sources of response error. There is less evidence on the extent to which people deliberately distort reports of their viewing or listening behavior. Most Americans seem to have a sense of what ratings data are, and how they can affect programming decisions. Again, anecdotal evidence suggests that some people view their participation in a ratings sample as an opportunity to "vote" for deserving programs, whether they are actually in the audience or not. While diary data may be more susceptible to such distortions than other methods, instances of deliberate deception, although real, are probably limited in scope.

A more serious problem with diary-based measurement techniques has emerged in recent years. As we noted earlier, the television viewing environment has become increasingly complex. Most homes now subscribe to cable and/or have a VCR attached to their set. In addition, remote control devices have become commonplace, as have small highly portable sets. These technological changes make the job of keeping an accurate diary more burdensome than ever. A viewer who has flipped through 20 channels to find something of interest may not know the channel to which he or she is tuned. Even if they record the channel indicated by the set, they may be in error because cable systems often change the channel designation of an over-the-air station. Nielsen lists cable systems and channel numbers in the diary. For reasons such as these, it is generally acknowledged that diaries under-report the audience for most cable networks and independent television stations. Other measurement techniques, however, can be used to compensate for many of these shortcomings.

There is also concern among TV broadcasters that changing lifestyles and the increased availability of portable sets has led to a significant amount of "out-of-home" television viewing. The traditional household diary has trouble measuring this use of the medium. The industry, therefore, has expressed some interest in exploring the use of "personal diaries" in television measurement, similar to those now used in radio.

Household meters have been the most important alternative to diary-based audience measurement. The best known metering device is Nielsen's Audimeter. The original Audimeter recorded radio listening, and required Nielsen field representatives to go to the homes equipped with these devices to retrieve their contents. Later, the record of radio or TV tuning recorded on motion picture film was mailed back to the Nielsen office, then in Chicago. Today, meters are a good deal more sophisticated, and are used only to record TV usage and channel tuning.

Modern meters are essentially small computers that are attached to all of the television sets in a home. They perform a number of functions, the most important of which is monitoring set activity. The meter records when the set is on and the channel to which it is tuned. This information is typically stored in a separate unit that is hidden in some unobtrusive location. Once or twice a day, the data it contains is retrieved from memory over a telephone line by a large computer.

For years, that was the scope of metering activity. And as such, it had enormous advantages over diary measurement. It eliminated much of the human error inherent in diary keeping. Viewing was recorded as it occurred. Even exposure of brief duration could be accurately recorded. Members of the sample did not have to be literate. In fact, they did not have to do anything at all, so no fatigue factor entered the picture. Because information was electronically recorded, it could also be collected and processed much more rapidly than paper-and-pencil diaries. Reports on yesterday's program audiences, called the "overnights" could be delivered.

There were only two major shortcomings to this sort of metering. First, it was expensive. It cost a lot to manufacture, install, and maintain the hardware necessary to make such a system work. That is still true today. As a practical matter, this means that metered measurement is viable only in relatively large media markets (i.e., nationally or in large urban areas). Second, household meters could provide no information on who was watching, save for what could be inferred from general household characteristics. The need to provide "people information," which is so essential to advertisers, has caused dramatic changes in how meters now function.

Peoplemeters had been under development in the United States and abroad for some time, but in the Fall of 1987 Nielsen began using them to generate national network ratings. Peoplemeters do everything that conventional household meters do, and more. With this type of metering, every member of the sample household is assigned a number that corresponds to a push button on the metering device. When a person begins viewing, they are supposed to press their button, thereby indicating their presence to the meter. When a person stops viewing, they are expected to press their button again. When the channel is changed, a light on the meter flashes until viewers reaffirm their presence (see Fig. 5.6). A system still only being used experimentally by Arbitron flashes an on-screen request (in the form of a "?") for viewer information, at specified intervals, even if the channel is not changed. All systems have hand-held units, about the size of a pack of cigarettes, that allow people to button push from some remote location in the room.

As with conventional meters, data are retrieved over telephone lines. At that point, all the button pushing and set-tuning activity can be com-

FIG. 5.6. Nielsen peoplemeter (reprinted with permission of Nielsen Media Research).

bined with data stored in a central computer to create people ratings. The introduction of peoplemeters triggered a storm of controversy about the method of measurement, and the samples on which it was based. As a relative newcomer, the merits and biases of this measurement technique are in somewhat greater doubt than more established techniques. Nevertheless, a number of generalizations and concerns seem warranted. These can, again, be categorized as issues of nonresponse and response error.

As is the case with diaries, a great many people who are sampled refuse to accept a peoplemeter. Both Nielsen and AGB, while it was in operation, experienced initial acceptance rates on the order of 50% to 60%. As always, the question is, "Are those who participate systematically different from those who do not?" For example, Although peoplemeters do not impose a formal literacy requirement, some have speculated that there is a kind of technological literacy required of respondents. The broadcast networks, which have seen their audience shares decline with the introduction of peoplemeters, have also criticized peoplemeter samples of over-representing those who subscribe to cable services. Moreover, lapses in button pushing and hardware failures reduce the effective in-tab samples on a day-to-day basis.

There are a number of response errors that seem to be associated with peoplemeters as well. Most notably, peoplemeters are believed to underrepresent the viewing of children. Youngsters, it seems, are not terribly conscientious button-pushers. More generally, there is concern about button pushing fatigue. How, for example, does one interpret instances in which the set is on but no one is reported watching? Conventional meters used to stay in households for 5 years. Doubts about the long-term diligence of peoplemetered homes, as well as pressure from the television networks, have caused Nielsen to turnover these households after only 2 years. Even so, some critics still believe that current methods of peoplemetering are fatally flawed.

Many of these problems could be solved if, like the old meters, peoplemeters required no effort on the part of respondents. The ideal device would be unobtrusive, yet capable of detecting specific individuals within the room.

These devices, called *passive peoplemeters,* are being developed. One of three technologies is likely to be employed. Infrared sensors will pick up heat sources, like human beings, in the room. The problem has been discriminating between different individuals, or for that matter, between dogs and children. As an alternative, sonic sensing devices could detect movement in the room. Here again, discrimination is a problem. How do you distinguish between a moving person and a curtain blowing in the breeze?

At present, the most promising technology for creating a passive peoplemeter is a computerized "image recognition" system. One such system, being developed by Nielsen, translates a person's image into a set of distinguishing features that it stores in a computerized memory. The system scans a pre-defined visual field and compares the objects it encounters with its memory to identify family members or visitors. Pictures of viewers, per se, are not stored or reported, only the incidence of recognized images.

Two other features of peoplemeter technology are worth mentioning here. First, peoplemeters are capable of monitoring VCR use. This is an important attribute, because two in three American households now own one. The system introduced by AGB worked by "fingerprinting" a tape as it was recorded. The fingerprint was an electronic code, laid down on an unused portion of video signal, that noted the date, and the channel being recorded. The fingerprint also imposed a running clock on the tape. That way, when the tape is replayed, the meter could determine when the program originally aired, and which sections of the show were played in fastforward. The latter information is of special importance to advertisers, because many people "zip" or "zap" commercials when they replay a program.

The second major addition to peoplemeters is a light pen capable of reading universal product codes (UPC). UPCs are the bar codes found on virtually all consumer products. In homes equipped with these "magic wands," respondents are asked to record their purchases by running the wand over the UPC of products they bring home from the store. The wand is then returned to a special cradle next to the peoplemeter, through which the information it contains is retrieved. Combining product purchase data with peoplemeter viewing data creates a vast store of information the industry refers to as *single-source* data.

In principle, single-source data are very appealing to advertisers, because they allow the user to describe an audience in terms of product purchases, rather than demographics. Because this allows an advertiser to zero in on its target market, more effective media buying could result. Arbitron has such a system it calls "ScanAmerica." Although it touts the advantages of its new "buyergraphics," others have adopted a wait-and-see attitude. The basic concern of critics is that adding wand waving to the button pushing tasks of peoplemeter samples will prove too burdensome, and greatly reduce cooperation rates. Nielsen also uses this wand technology with a large panel of household, but it does not produce its ratings report with that sample.

Telephone questioning, of course, is the oldest formal method of data collection used by ratings services. It is still used today, and in some forms, is considered the standard against which all other methods of measurement are to be judged. Data collection over the telephone takes one of two form: recall or coincidental.

Telephone recall, as the name implies, requires a respondent to remember what they have seen or heard over some period of time. Generally speaking, the quality of recalled information is affected by two things. One is how far back a person is required to remember. Obviously, the further removed something is from the present, the more it is subject to "memory error." Second, is the salience of the behavior in question. Important, or regular occurrences are better remembered that trivial or sporadic events. Because most people's radio listening tends to be regular and involve only one or two stations, it is believed that the medium's use can be accurately studied with telephone recall techniques.

Birch Radio uses telephones to produce local market reports in direct competition with Arbitron, whereas SRI's RADAR provides estimates of national radio usage and network audiences. Although there are differences in their methods, both companies call a random sample of listeners, and ask them to report on recent radio listening. Interviewers ask questions that identify listening at specific times within specific dayparts. If a respondent does not know a station's call letters, other identifying information like a station's frequency or slogan can be used. Birch speaks to each respondent only once, and asks about the prior day's listening. RADAR, on the other hand, interviews a person once each day for a week, and asks about radio use from the time of the previous contact to the present.

Telephone recall techniques have a number of advantages compared to the major alternative, radio diaries. First, telephone interviewing achieves higher levels of cooperation. Birch and RADAR report response rates on the order of 55% to 65%. Although people without telephones are, by definition, excluded, overall higher response rates reduce the likelihood of nonresponse error. Second, because respondents are verbally questioned, there is no literacy bias in the method. If a Hispanic household is sampled, a Spanish-speaking interviewer can be employed. Third, because the research firm takes the initiative by calling respondents each day, there is no end of the week diary fatigue. Fourth, telephone techniques work particularly well for gathering data from younger listeners who tend to be poor diary keepers.

Like all other methods of data collection, however, telephone recall has certain limitations. If people are only questioned about their previous day's listening, week-long patterns of audience accumulation can only be inferred from mathematical models. We talk more about modeling in the last chapter. The use of interviewers can also introduce error. Although interviewers are usually trained and monitored in centralized telephone centers, they can make inappropriate comments or other errors that bias results. Finally, the entire method is no better than a respondent's memory. Even though people are only expected to recall yesterday's listening, there is no guarantee that they can accurately do so.

As C.E. Hooper argued some 50 years ago, *telephone coincidentals* can offer a way to overcome problems of memory. These surveys work very much like telephone recall techniques, except that they ask respondents to report what they are seeing or listening to at the moment of the call. Because respondents can verify exactly who is using what media at the time, errors of memory and reporting fatigue are eliminated. For these reasons, telephone coincidentals are widely regarded as the standard against which other methods of measurement should be evaluated. Most new measurement techniques, therefore, are obliged to offer a comparison of their results with a concurrently executed telephone coincidental.

Despite this acknowledged superiority, no major ratings company routinely conducts telephone coincidental research. There are two problems with coincidentals that militate against their regular use. First, a coincidental interview only captures a glimpse of a person's media use. In effect, it sacrifices quantity of information for quality. As a result, to describe audiences hour to hour, day to day, and week to week, huge numbers of people would have to be called around the clock. That becomes a very expensive proposition. Second, as with all telephone interviews, there are practical limitations on where and when calls can be made. Much radio listening occurs in cars (without cellular phones). Much television viewing occurs late at night. These behaviors cannot be captured with strictly coincidental techniques. For these, and other reasons, the coincidental telephone method is no longer used for any regular rating service.

There are other methods of measurement that could conceivably be used. One possibility is to monitor television set use on specially designed cable systems. Another is to scan the airwaves with radar-like devices that determine how nearby sets are tuned. Some researchers have suggested replacing conventional diaries with calculator-like "electronic diaries." Each of these has certain appeals and significant drawbacks that we do not delve into here because none is currently used by a major U.S. producer of ratings research.

PRODUCTION

Issues of sampling and measurement are well known to survey researchers, and there are large bodies of academic literature offering research and theory on these topics. We have, therefore, some well-established criteria by which to judge the work of the ratings services. But sampling and measurement alone do not a ratings book make. The data collected by these methods must undergo a production process, just as other raw materials are turned into products. Here, standards of what is or is not appropriate are harder to come by. Yet, no discussion of ratings methods would be complete without mention of the production process. Every ratings company does things a little differently, but basically, the production process involves three activities: editing the data, adding new information, and making projections.

Editing

Ratings companies are continually flooded with data that must be digested and turned into a useful product. One of the most difficult sources of data to deal with is the diary. Hundreds of thousands of hand-written diaries arrive at Arbitron and Nielsen each year. They must be checked for accuracy, logical inconsistencies, and omissions. They must also be translated into a form that a computer can deal with. The process of getting clean, accurate, complete data ready to be processed is called *editing*. It can be a very laborious activity, and despite serious efforts at quality control, it is here that *processing error* is most likely to be introduced into the ratings.

Diary editing involves a number of activities that are performed by either people or machines. First, it must be determined that a returned diary is usable. It might, for example, have been filled out in the wrong week, mailed too late to be useful, or simply be too incomplete for inclusion. It must be checked for logical inconsistencies. Suppose a radio diary indicates that a listener heard station "KREO" in his or her car, but Arbitron knows that no such station is receivable in that market? Or suppose that a television diary reports someone watched a program, but it has the wrong channel number or call letters associated with it? Strict editing procedures will usually prescribe a way to resolve these discrepancies.

Suppose, however, that information is just plain missing? Rather than throw out an otherwise usable diary, ratings companies will often "fill in the blanks" through a process called *ascription*. These procedures typically use computer routines to determine the answer with the highest probability of correctly filling that blank. For example, if Nielsen receives a diary with the age of the male head of household (e.g., 31), but not the age of the female head of household, it consults age and gender tables, and "guesses" that her age would be 3 years less than her husband's (i.e., 28). Analogous ascription techniques are used to determine the identity of stations heard, or the duration of media use if such data are missing. While these practices strike some as questionable or improper, ascription is a standard procedure in virtually all survey work, and is typically based on systematic methodological research.

Editing can also involve definitional questions. Take, for instance, the data recorded by a meter. If a person watches less than half a program, should they nonetheless be included in the program's total audience? The standard practice in television viewing has been to credit one quarter-hour of viewing to a program if at least 5 minutes of use has taken place. Under that definition, of course, a person might show up in more than one program audience in a given quarter-hour. Similarly, RADAR will credit a listener to a radio network if he or she heard the radio for at least 3 minutes in a quarter-hour period.

Often times there is no clear right or wrong answer to such definitional questions. It is more a matter of what the industry will agree to accept. As media and measurement technologies change, new questions arise, and new solutions must be negotiated by the parties at interest. For example, if a household watches one program, but tapes a second one on the VCR, should that household be credited to the second program's audience? At present, the answer is yes. The ratings services treat that household as if it had viewed the program at the time it aired. Obviously, there are other ways to credit the audience for the taped program. The resolution of these definitional questions is often arbitrary. If, however, some party feels disadvantaged by a particular editing procedure, it may become the subject of a political struggle within the industry.

Programs, Schedule, and Other Information

Despite the vast amounts of information collected by diaries, meters, and telephone calls, these data are not, in and of themselves, sufficient to produce audience ratings. Other information must be added to make a complete, usable product. The most important addition to data on people's set-tuning behavior is information about the programming on those sets. Station schedules are needed to check the accuracy of diary entries. Further, even the most sophisticated of meters are not capable of determining what program was on which channel at what time. These data must, somehow, be collected and added to the ratings database.

Because radio listening is credited to stations, rather than to specific programs, the problem is relatively simple. Arbitron, for example, mails radio stations an "information packet," in which stations verify their call letters and report their network affiliations, broadcast schedules, and especially current slogans, catch phrases, and station identifications—"96 Rock," "News Radio 88," "All News 67," "Continuous Country," "Z 104," "Y95," "My Kind of Country," "Lazer 103," and "98 FM." There are frequent arguments, and occasional lawsuits, over who is entitled to phrases such as "More Music" or "Music Radio." Further, if two stations in nearby markets are at 102.7 and 103.1 there may be confusion if both use "one-o-three" in their phrase. Station personnel sometimes travel to the research companies, or hire consultant firms to check for that, in the hope of examining the diaries and finding uncredited listeners.

Television viewing, on the other hand, must be associated with every specific programs. Much more detailed information is needed. Ratings companies usually get this by having stations fill out "program title logs." These require the station to report the programs airing in every quarter hour, of every broadcast day, for every day of the week, across all survey weeks. Handling program schedules like these would be problem enough, but the growth of new television technologies has expanded the problem to nightmarish proportions. Cable television has greatly complicated the task of determining what is on which channels. There are 10,000 cable systems in the United States. A majority of these have over 30 channels of programming, including dozens of cable networks, access channels, and local stations—the latter sometimes from several different TV market areas. Different cable systems can, and do, carry these services on different channels. Even local TV stations may be "remodulated" to a new channel number. In any given television market area, there may be dozens of such cable systems—frequently using different channel assignments. Figure 5.7 is a Cable Conversion Chart from the *Washington Post's* TV listings section. We have included it here to give you a sense of how hard it can be, even in one market, to tell what is on any given channel. Now if you can imagine that situation repeated in various markets around the country, you will have some idea of what confronts a national TV ratings service.

Nielsen, in fact, has a special service called the Cable On-Line Data Exchange (CODE) that tracks what is carried on every channel of every cable system in the United States. This information needs to be matched with its peoplemeter data to determine cable network audiences.

The job of figuring out what is on TV would be easier if each program contained a "signature" that a machine could simply read. In fact, such a system does exist. The broadcast networks have, for some years, cooperated with Nielsen by imposing a special electronic code in the video portion of their broadcast signal. This system, called the Automated Measurement of Lineups (AMOL) allows detection devices in each market to determine when affiliates are broadcasting a network program. Unfortunately, not all programs (e.g., local productions, PBS programs, some network reruns and some syndicated) contain such an electronic code, so more traditional techniques must still be employed.

In addition to programming information, other data enter into the production of ratings reports as well. For example, stations occasionally have technical difficulties. These may affect their audience ratings, so they are reported in the ratings books. Stations may also engage in extraordinary activities to boost their ratings during a sweeps period. The ratings services keep an eye out for any "special station activities" intended to bias or distort the ratings, because it is thought to compromise the integrity of the entire process. Depending on the transgression, the ratings companies will either note the offending station's crime in the ratings book, or drop the station's ratings from the book altogether.

Projections

Ultimately, the ratings services must publish their estimates of audience size and composition. This process uses sample information to make a projection of what is true for the entire population. Suppose, for example, we used a sample of 1,000 individuals to study a population of 1 million. In effect, that would mean that each member of the sample represented 1,000 people in the population. If 50 people in our sample watched a local news show, we could project the show's actual audience to be 50,000. That is essentially what the ratings services do. They determine the number

CABLE CONVERSION CHART

DISTRICT | **MARYLAND** | **VIRGINIA**

District (Cablevision)	Annapolis (United Cable)	Anne Arundel (Jones Intercable)	Baltimore County (Comcast)	Calvert (Jones Intercable)	Charles (Jones Intercable)	Columbia (Howard Cable)	Frederick (Cablevision)	Gaithersburg (Cablevision)	Lexington Park (United Cable)	Montgomery (CableTV Mont.) Red System / Blue System	Prince George's (MetroVision)	Prince George's (Prime Cable)	Station	Alexandria (Jones Intercable)	Arlington (Cable TV Arlington)	Dale City-W'bridge (CV of PW)	Fairfax (Media General)	Fort Belvoir (Columbia Cable)	Leesburg (Prime Cable)	Loudoun (Prime Cable)	Manassas (Cablevision)	Pr. William/Lake Ridge CV	Reston (WCC)
D.C. Area Broadcast																							
4	4	4	—	4	4	4	4	4	4	21/4	A4	B4	4	4	4	4	4	3	4	4	4	4	2
5	5	5	25	5	5	5	5	5	5	5/5	A5	B5	5	5	5	5	5	6	5	5	5	5	6
7	7	7	—	7	7	7	7	7	7	24/7	A7	B7	7	7	7	7	7	8	7	7	7	7	10
9	9	9	—	9	9	9	9	9	9	26/9	A9	B9	9	9	9	9	9	10	9	9	9	9	11
20	28	3	30	20	6	20	6	3	A10	65/40	A20	B20	20	20	20	12	20	12	11	2	20	6	12
26	26	6	3	18	3	33	8	8	12	23/26	A8	B26	26	26	26	10	26	20	10	3	26	19	3
32	32	21	39	19	—	32	24	6	—	29/32	A10	B19	32	32	38	36	32	32	6	6	32	20	24
23	—	—	—	42	31	—	—	—	—	85/50	A28	B23	50	35	50	24	50	—	48	50	52	40	28
—	—	—	—	—	—	—	—	—	—	—/—	—	B24	53	19	10	34	53	11	8	12	53	21	13
—	—	—	—	21	—	—	36	—	—	—/—	A31	B21	56	34	54	—	56	25	32	11	56	—	23
—	—	—	—	—	—	—	—	12	—	—/—	A33	—	66	37	—	19	52	22	19	10	60	22	16
Baltimore Area Broadcast																							
44	2	2	12	2	36	2	2	2	3	2/2	A2	B2	2	28	2	—	41/4	9	41	—	2	16	17
45	11	11	21	11	35	11	11	11	11	11/11	A11	B11	11	29	11	—	11	16	42	—	11	17	18
46	13	13	23	13	33	13	13	13	13	13/13	A13	B13	13	30	13	—	13	7	43	—	13	18	20
33	22	12	22	22	2	29	3	—	—	19/22	A16	B22	22/67	22	22	—	22	13	12	8	22	—	14
—	—	—	—	—	—	6	—	—	—	—/—	B14	—	24	—	—	—	—	—	—	—	—	—	—
24	30	10	14	16	—	30	12	10	A8	80/45	A23	B27	45	45	45	—	45	30	44	—	—	—	15
—	35	22	33	—	—	35	—	—	—	—/—	A15	B25	54	—	—	—	54	—	—	—	—	—	—
Premium Cable																							
12	33	14	—	29	24	—	—	—	—	—/—	—	A12	American Movie Classics	2	—	16	88	—	—	—	43	39	—
49	—	—	—	—	—	—	—	—	—	17/20	—	B30	BRAVO	42	49	—	71	—	—	—	42	—	—
47	—	32	8	24	19	—	14	L	7	14/17	—	B33	Cinemax	31	18	14	76	—	21	42	29	35	—
10	18	33	—	41	14	25	15	—	6	115/16	—	B35	Disney Channel	40	21	18	65	23	40	45	30	32	—
—	—	—	—	—	—	—	—	—	—	—/—	—	B36	Galavision	—	—	—	115	—	—	—	—	—	—
3	16	35	5	38	17	22	16	H	2	113/14	A21	B29	Home Box Office	8	16	21	74	19	18	41	28	2	B10
11	20	34	17	23	32	26	19	—	12	16/19	B2	A24	Home Team Sports	38	6	20	86	—	—	47	41	37	—
48	21	30	—	40	21	18	—	—	—	15/18	A19	B34	The Movie Channel	—	—	22	81	24	20	43	46	34	B8
—	—	—	2	37	—	—	—	—	—	—/—	—	B37	Playboy Channel	—	—	—	—	31	—	—	—	—	—
2	—	18	20	39	—	23	—	—	—	114/15	A17	B32	Showtime	23	14	15	67	2	17	44	44	33	27
Non-Premium Cable																							
27	23	29	35	31	23	36	32	—	—	1/36	B13	A28	Arts & Entertainment	16	35	23	70	—	47	34	48	31	B13
—	—	14	—	—	—	—	—	—	—	—/—	—	A5	Associated Press	—	—	—	108	—	—	—	—	—	19
—	—	—	—	—	—	—	—	—	—	—/—	—	A5	AP Business Plus	—	—	—	107	—	—	—	—	—	—
6	33	—	—	6	15	—	—	—	—	7/27	B16	A13	Black Entertainment	47	41	—	57	—	—	30	24	—	—
13	14	27	28	28	34	12	26	J	10	10/10	B11	A14	Cable News Network	6	12	2	80	21	29	24	8	24	B11
38	24	38	11	33	25	31	35	D	—	28/31	B3	A35	CBN Cable Network	44	23	28	69	26	27	26	6	23	—
30	34	28	10	61	29	14	17	—	—	66/41	B12	A15	CNN Headline News	24	15	31	82	33	33	35	10	8	25
15	31	19	18	27	—	16	34	E	—	82/47	B9	A19	C-Span	36	52	33	17	—	45	21	25	26	26
14	20	40	37	—	—	—	—	—	—	83/48	—	A20	C-Span II	27	53	—	19	—	—	—	—	—	—
—	—	—	—	—	—	—	—	—	—	—/—	—	—	Country Music Television	—	—	—	111	—	—	—	—	—	—
17	6	20	38	34	22	—	40	—	—	69/44	B10	A33	Discovery	43	34	30	96	5	16	32	47	—	—
25	10	26	16	25	10	3	21	G	B8	6/6	B17	A32	ESPN	12	24	6	85	36	22	23	34	13	B2
28	—	—	15	37	8	36	—	—	—	31/34	B24	A9	Financial News Network	33	61	16	99	15	34	27	38	—	—
—	—	—	—	—	—	—	—	—	—	—/—	—	—	Genesis	—	—	—	113	—	—	—	—	—	—
—	—	15	—	15	16	21	—	—	—	—/—	A30	—	Home Shopping Network	49	—	—	83	—	39	15	27	—	—
18	—	—	—	34	—	—	—	—	—	81/46	—	A3	Learning Channel	—	34	—	97	—	35	18	49	36	—
—	—	—	—	—	—	—	—	—	—	—/—	—	—	Liberty Network	—	—	—	116	—	—	—	—	—	—
37	25	16	24	32	18	34	27	—	—	40/37	B8	A29	Lifetime	14	46	26	72	35	28	20	15	25	22
16	—	8	10	10	15	15	29	—	—	70/55	B9	B10	Local Government	34	35	—	16	4	—	—	1	—	—
43	12	31	34	35	13	10	18	R	B11	8/8	B16	B6	Music TV	13	3	13	77	28	38	28	33	14	B3
19	23	25	19	30	11	28	22	F	B13	68/43	B4	A17	Nickelodeon	15	48	11	64	27	2	33	16	27	B13
—	—	—	—	—	—	—	39	—	—	—/—	A36	A31	PTL	—	—	32	117	—	30	—	18	—	—
—	3	—	26	—	—	—	—	28	—	25/28	B23	A21	QVC—Shopping	—	27	—	84	—	—	—	—	—	—
—	—	—	—	—	—	—	—	—	—	—/—	—	A6	Radar Weather	—	—	—	102	—	—	—	—	—	—
—	—	—	—	—	—	—	—	—	—	—/—	—	A4	Sports Tracker	—	—	—	—	—	—	—	57	—	—
—	—	—	—	—	—	—	—	—	—	—/—	—	A2	TEMPO	—	—	—	98	14	36	19	54	29	—
41	27	36	27	36	20	—	25	—	—	27/30	B7	A26	The Nashville Network	—	40	27	109	—	31	31	23	15	—
—	—	—	—	—	—	—	—	—	—	67/42	—	—	Trinity Broadcasting Network	—	—	—	114	—	—	—	—	—	—
31	—	—	—	—	—	—	—	—	—	9/29	A32	A27	Univision—SIN	48	51	—	58	—	—	—	—	—	—
42	—	—	29	—	12	—	38	—	—	30/33	B20	A10	Video Hits-1	46	39	29	110	—	46	29	39	—	—
8	8	23	9	26	26	27	20	K	B9	20/23	B6	A34	USA Network	11	25	25	75	17	3	25	31	11	B6
—	19	—	—	—	27	—	—	—	—	—/—	A25	—	WGN, Chicago	—	—	—	89	—	26	17	—	—	—
22	—	24	32	—	30	19	—	I	—	22/25	A34	B28	WWOR, New York	25	19	17	79	29	25	—	21	12	—
21	17	17	3	17	28	17	23	C	A6	12/12	A35	B8	TBS, Atlanta	17	17	8	87	18	24	16	17	10	21
36	29	39	13	8	—	24	33	—	—	39/35	B21	A7	Weather Channel	21	36	35	101	34	23	22	14	28	—

FIG. 5.7. Cable conversion chart (compliments of *The Washington Post;* reprinted from Salvaggio & Bryant, 1989).

of people represented by one in-tab diary and assign that diary an appropriate number. If people are the unit of analysis, the number is called *persons per diary value* (PPDV). If households are the unit of analysis the number is labeled HPDV (i.e., households per diary value).

This illustration works quite well if we have perfect probability samples, in which all members of the population are proportionately represented. As we have seen, however, that is never the case. Nonresponse error means that some kinds of people are over-represented, whereas others are under-represented. Remember, also, that the most common remedy for this problem is to weight the responses of some sample members more heavily than others. Suppose, in the illustration just given, that 18- to 24-year-old males were under-represented in the in-tab sample. Let's say they constitute 4% of the sample, but are believed to be 8% of the population. Males in this group would receive a weight of 2.0 (i.e., 8% / 4% = 2.0). Therefore, to project total audience size for this group, each young man should have a PPDV of 2,000 (i.e., 1,000 x 2.0), instead of 1,000. Conversely, over-represented groups should have PPDVs of less than 1,000.

In practice, the weights that are assigned to different groups are rarely so extreme as the illustration just given (i.e., they come closer to 1.0). Further, ratings services will weight a single respondent on a number of variables besides age and gender to make a final determination of PPDVs. Although this method of audience projection is not without biases, it is generally agreed that it is the best practical remedy for nonresponse errors.

Could similar, statistical solutions correct for some of the measurement errors we reviewed in the preceding section? After all, we have noted that certain kinds of response errors are associated with certain kinds of measurement. Some work in this area has been done, but there is less consensus on how such statistical corrections should be applied to formal published audience estimates.

The best illustration of this problem occurs in reconciling meter- and diary-based estimates of television audiences. Prior to the introduction of peoplemeters, Nielsen had to use both household meters and diaries to estimate national network audiences. Somehow, these data had to be integrated into a single "best guess" as to audience size and composition. Because metered data were assumed to more accurately measure set usage, they were used to fix audience size, whereas diary data (which often showed smaller audiences) were extrapolated to determine the likeliest demographic breakdown. At this writing, the same situation now exists in major local markets that are measured with both diaries and conventional household meters.

But what of smaller markets that are only measured with diaries? This method is generally agreed to under-represent the audience for independent television stations. Here too, a statistical formula to correct for systematic error could be employed. This procedure is called *calibration,* which in effect, adjusts the ratings of stations to the levels that would be expected under a system of metered measurement. Although independent stations are, understandably, enthusiastic about this procedure, the ratings companies have been reluctant to introduce calibration into the production process. This reluctance is due, in part, to the resistance of affiliated stations that would be disadvantaged by calibration. These stations are, not insignificantly, major clients of the ratings companies.

Here again, we can see how industry politics play a role in the methods used to estimate audiences. As long as major measurement systems, each with its own biases, are used side by side, questions of calibration will continue. Whether these, or other changes, work their way into ratings data has yet to be determined. In any event, a ratings analyst should know the consequences of these methodological issues when they use the services of ratings companies.

RELATED READINGS

Anderson, J. A. (1987). *Communication research: Issues and methods.* New York: McGraw-Hill.
Babbie, E. (1983). *The practice of social research* (3rd ed.). New York: Wadsworth.
Beville, H. M. (1988). *Audience ratings: Radio, television, cable* (rev. ed.). Hillsdale, NJ: Lawrence Erlbaum Associates.
Fletcher, A. D., & Bower, T. A. (1988). *Fundamentals of advertising research* (3rd ed.). Belmont, CA: Wadsworth.
Webster, J. G. (1983). *Audience research.* Washington, DC: National Association of Broadcasters.
Wimmer, R. D., & Dominick, J. R. (1991). *Mass media research: An introduction* (3rd ed.). Belmont, CA: Wadsworth.

RATINGS RESEARCH PRODUCTS 6

Up to this point we have concentrated on the evolution of ratings firms, and how they go about collecting and processing data. Like any commercial enterprise, however, they must produce goods or services that can be sold in the marketplace. In this chapter, we will consider the products that ratings companies offer for sale.

The most influential consumers of ratings data are those who buy and sell time. These include people in network and station sales, station representatives, advertisers, and advertising agencies. Although other users of ratings data are certainly important, this first group is critical in terms of product development. For that reason, it makes sense to organize ratings products by the advertising markets they are intended to serve. As described in chapter 1, the major markets are network, local, and syndication.

There are a great many reports and services offered by ratings companies. The enormous databases that these firms collect allow them to create far more products than we can possibility review here. Furthermore, the number of products is on the rise as ratings data are combined with other sources of information, and computers open up new ways to manipulate, merge, and present the data.

In light of these considerations, our description of ratings products will provide only selected examples of the better known and more widely used reports and services. By concentrating on these we can acquaint the reader with the most common report formats, and demonstrate how some of the research concepts we introduced in the previous chapter pop up in the context of an actual ratings report. We leave it to our readers to explore for themselves the many variations on a theme that the ratings services offer.

NETWORK RATINGS

Network television ratings are certainly the most visible of all the ratings products. Indeed, for most Americans, the Nielsen name has become synonymous with ratings. That identification occurs for good reason. Since the demise of C.E. Hooper, the Nielsen company has dominated national ratings. It is the *Nielsens* that are often held to account for the cancellation or renewal of network television programs—an explanation that belies the complexity of programming decisions. The Nielsen service that actually provides network ratings is called the Nielsen Television Index (NTI).

Today, NTI bases all of its network ratings on a single sample of households equipped with peoplemeters. There are approximately 4,000 of these households in the Nielsen sample, or, with an average of about 2.5 people in each home, roughly 10,000 individuals. At any given point in time, however, the actual number of households providing useful data will be less.

Nielsen selects households for inclusion in the sample through a procedure of multi-stage area probability sampling. As we described in the previous section, this process works its way through a series of sampling units. Nielsen first samples counties, then census blocks, then city blocks, and finally households. In the last stage, Nielsen must secure the cooperation of each designated household. Usually, just over 50% agree to cooperate. If a predesignated household refuses to accept a meter, Nielsen substitutes another "matched" household. This rule of substitution means that the Nielsen sample deviates from the strict definition of a probability sample. For all intents and purposes, however, both Nielsen and its customers treat it like a random sample. To keep the sample fresh, Nielsen replaces each household in the sample after 2 years.

Every television set in a sampled household is connected to a peoplemeter, which collects information on set tuning and viewers for a period of 2 years. These data are retrieved over telephone lines by Nielsen computers. Before looking at the reports Nielsen publishes, it is worth reflecting on the enormous amount of data this system generates. Ten thousand people watching various combinations of broadcast television, VCRs, and cable, being monitored minute by minute over a period of years, creates a vast flow of raw material to be processed into useful reports and services.

The best known, and longest continuously produced, television network ratings report is NTI's National TV Ratings, better known as the "pocketpiece." So named for its small vest pocket size, the pocketpiece is issued

RATINGS RESEARCH PRODUCTS

once a week and provides a variety of the most commonly used audience estimates.

Figure 6.1 shows two facing pages out of a pocketpiece report. In this section of the book, NTI displays the television household (TVHH) ratings for prime-time network programs in a way that highlights the scheduling characteristics of those programs. These pages depict the ratings for a Thursday night in January. Across the top of the page is a banner indicating the time periods, in quarter hours, and the HUT level associated with each time period. Note that the HUT level was highest between 9:15 and 9:30. During that time, Nielsen estimated that 67% of TVHH had a set in use. Because network programs run at different times in different time zones, Nielsen adjusts its audience estimates to the Eastern time zone.

Down the left-hand side of these pages are the various networks, or station categories, that households are likely to be watching. On the upper page are the three major broadcast networks—ABC, CBS, and NBC. Audiences for Fox are reported elsewhere in the pocketpiece. On the lower page are the estimated audiences for independents, cable networks, and public television. As you can see, no specific program audience information is presented for this latter class of program services.

On this particular Thursday evening, the network prime-time schedule began with ABC showing "Knightwatch," CBS showing "48 Hours," and NBC showing the "The Cosby Show." The first number under each program title is the average audience for that program. It is expressed as the total projected number of households watching in an average minute. Just under that the same number is a percentage of total TVHH, which of course, is the program's rating. For example, "The Cosby Show" was viewed by 26,940,000 households in an average minute, which means it has a TVHH rating of 29.8. The number beneath the program rating is the program's share. In this case, "Cosby" was being watched by 45% of all the households using television at that time. The last number in a column is the average audience in a specific quarter hour.

By arranging program audience estimates in this way, Nielsen gives the reader a clear sense of how different networks and non-network services do in competition for the available audience. It also suggests something about audience flow from one quarter hour to the next, although bona fide analyses of audience flow require access to different data. These tables do not tell us anything about the demographic composition of program audiences. The pocketpiece reports that information in a different section of the book. Nielsen arranges network program audience estimates both alphabetically and by time period. In these sections individual program audiences are broken down into 20 different age and gender combinations, including categories for "working women" and "LOH W/CH <3" (i.e., lady of house with child less than 3). The precise demographic catego-

Nielsen NATIONAL TV AUDIENCE ESTIMATES

EVE. THU. JAN. 12, 1989

TIME	7:00	7:15	7:30	7:45	8:00	8:15	8:30	8:45	9:00	9:15	9:30	9:45	10:00	10:15	10:30	10:45
HUT	62.4	63.6	63.8	65.5	66.7	65.9	66.3	66.5	67.0	65.8	65.7	64.1	62.9	61.4	59.0

ABC TV

←――――――――――KNIGHTWATCH――――――――――→ ←――――――――DYNASTY (PAE)――――――――→ ←――――――――HEARTBEAT――――――――→

AVERAGE AUDIENCE
(HHds (000) & %)   5,700   6.1 *   6.5   *6.10   9,760 *10.8   10.2 *   11.5 *7.0   6,330   6.7 *   7.4 *
SHARE AUDIENCE %   6.3   9   10   16   15   18   11   12
AVG. AUD. BY 1/4 HR 6.4   5.9   6.3   6.7   9.9   10.5   11.4   11.5   6.7   6.6   7.2   7.6

CBS TV

←――――――48 HOURS HIGH SCHOOL―――――→ ←――――――PARADISE (PAE)――――――→ ←――――KNOTS LANDING――――→

AVERAGE AUDIENCE
(HHds (000) & %)   9,850   10.6 *   11.2 *10.2   9,220 *10.2   9.7 *   10.8 *5.6   14,100   15.7 *   15.5 *
SHARE AUDIENCE %   10.9   16   17   15   15   18   25   26   26
AVG. AUD. BY 1/4 HR 10.8   10.3   11.2   11.1   9.7   9.7   10.4   11.2   15.5   15.8   15.6   15.3

NBC TV

BILL COSBY SHOW | A DIFFERENT WORLD | CHEERS | DEAR JOHN | ←――L.A. LAW――→

AVERAGE AUDIENCE
(HHds (000) & %)   26,940 10.6 *   23,590   22,960   18,890   19,160   21.3 *
                    29.8   26.1   25.4   20.9   22.2   35
SHARE AUDIENCE %   45   39   38   32   34   21.1
AVG. AUD. BY 1/4 HR 28.3   31.3   26.1   25.3   25.5   20.9   20.8 21.1   21.0

    11.9   12.4   11.7   11.9   11.6   10.3
    18   19   18   18   18   17

INDEPENDENTS (INCL. SUPERSTATIONS)

AVERAGE AUDIENCE   17.4   14.9
SHARE AUDIENCE %   28   23

SUPERSTATIONS

AVERAGE AUDIENCE   5.3   4.6   3.3   3.3   3.1   3.1   2.9   2.9
SHARE AUDIENCE %   8   7   5   5   5   5   5   5

PBS

AVERAGE AUDIENCE   2.1   2.5   2.9   3.1   3.2   3.1   2.3   1.9
SHARE AUDIENCE %   3   4   4   5   5   5   4   3

CABLE ORIG.

AVERAGE AUDIENCE   7.3   7.3   7.3   8.8   8.1   8.4   7.2   6.4
SHARE AUDIENCE %   12   11   11   13   12   13   11   11

PAY SERVICES

AVERAGE AUDIENCE   2.0   2.6   2.6   2.7   5.0   3.8   5.6   3.3
SHARE AUDIENCE %   3   4   4   4   8   6   9   5

U.S. TV HOUSEHOLDS: 90,400,000

For explanation of symbols, See page 8.

FIG. 6.1. NTI pocketpiece (reprinted with permission of Nielsen Television Index).

ries Nielsen will report varies by daypart. Nielsen will also report the percentage of TVHH that recorded a program with the set off or tuned to another program. Rarely does that VCR audience account for more than one rating point.

NIT offers a host of other published reports. These are described briefly in Appendix A. Among them are a variety of cable network ratings, provided through a division that Nielsen calls the Home Video Index. Nielsen also releases a number of special reports from time to time. These include things like: *Cable TV: A Status Report; VCR Tracking Report; Viewing to Political Telecasting;* and *Television Audience,* which is an annual compendium of audience data the company has issued since 1967.

Television, of course, is not the only advertiser-supported medium providing network service to the public. Although they cannot rival the amounts spent on television, radio networks still attract millions of advertiser dollars, and need ratings services. As noted in chapter 4, shortly after Nielsen ended its radio network measurement, a research effort called RADAR was initiated to fill the void. Today, Statistical Research Inc. (SRI) publishes the RADAR reports, which are the only true radio network ratings service.

RADAR ratings reports are based on telephone interviews conducted with a sample of 12,000 respondents. SRI determines which households will be called through a process of random digit dialing (RDD). Within each home, SRI randomly selects one individual age 12 or older, and interviews him or her once a day for the next week. Interviewing goes on for 48 weeks each year. SRI typically has response rates in excess of 65%.

RADAR reports are issued in three volumes, twice each year in the spring and fall. Each edition of the RADAR report includes information collected over the past 12 months. The first volume, entitled *Radio Usage,* contains general information about the composition and listening habits of the audience during different dayparts and quarter hours, without regard to specific networks. The measurements RADAR reports include the size of the audience in an average quarter hour (AQH), as well as 1-day, 5-day, and 7-day cume estimates. These audience summaries are broken out by standard age/gender groupings as well as other demographic (e.g., income and education), geographic, and behavioral variables.

Volumes 2 and 3 are entitled *Network Radio Audiences to All Commercials* and *Network Radio Audiences to Commercials Within Programs.* Basically, these are audience estimates for the 20 or so radio networks measured by RADAR. These estimates are made possible by combining program and commercial clearance data obtained from the networks with the station listening information obtained from the respondents. However, because some stations extract network commercials from network programs and air them separately, audience estimates are reported in two volumes. Volume 2 reports the total audience for network commercials, whether they are aired with the program or not. Volume 3 estimates do not include commercials outside the program. Needless to say, audience estimates in Volume 2 are greater than or equal to those reported in Volume 3.

In addition to its standard ratings reports, SRI offers more varied breakouts of the data through an on-line computer facility called RADAR On-Line (ROL), and through personal computers via reports on disks. SRI also does more specialized audience studies. The firm has a well-established reputation for doing quality work, and is often called on to do the telephone coincidentals against which other measurement techniques are evaluated.

National Public Radio also produces an estimate of the audiences for its programming but those ratings are based on Arbitron data from a sample of stations. From the reports of these stations, using weighing, AQH and cume figures are projected.

LOCAL RATINGS

Both radio and television are measured on a local market-by-market basis. Local ratings, however, are usually available from two different suppliers. In television, the Nielsen Station Index (NSI) and Arbitron are long time competitors. In radio, Arbitron, which used to have the field pretty much to itself, now faces competition from Birch Radio. We consider the local television ratings first.

There are many similarities in the research services offered by Arbitron and Nielsen. Certainly, some differences of method, product, and price do exist—differences that the salespeople for each service are likely to emphasize—but for our purposes, the similarities outweigh the differences. Indeed, the kinds of ratings data that are available to a TV station differ more by the size of its market than by the name of its supplier. Ratings research in larger markets is based on bigger samples, offers different measurement options, more services, and is much more expensive. It is no coincidence that larger markets also tend to be much richer in terms of the dollars spent on media. For these reasons, it is important to expand on our earlier discussion of local markets.

Both services organize the United States into roughly 215 mutually exclusive television market areas. Each market area is a collection of counties in which the preponderance of total viewing can be attributed to local or home-market stations. That is, counties are assigned to markets on the basis of what stations the people in those counties actually view. Each market area is called an Area of Dominant Influence (ADI) by

RATINGS RESEARCH PRODUCTS

Arbitron or a Designated Market Area (DMA) by Nielsen. Figure 6.2 is a map of all the DMAs in the country.

As you can see, market areas vary substantially in terms of their sheer geographic size. More importantly, however, they differ in terms of population. Population size is, in turn, used to rank markets from largest to smallest (See Appendix C for a ranking of ADIs). Of course, shifts in the U.S. population cause changes in how markets are ranked. But, because markets areas are ultimately defined by viewing behavior, changes in programming, transmitters, cable penetration, and so on can also alter market size and composition.

Every year the ratings services reconsider how markets should be constituted, and changes do occur. Sometimes, counties on the border between two adjacent markets will be moved from one to the other. Such changes are no small matter. On one hand, national spot buys are sometimes made in the "top 20" or "top 50" markets. If the loss of a county causes a market to drop below an important breakpoint, it can have a detrimental impact on every station in the market. On the other hand, redrawing market boundaries might have a differential impact on local stations. Because of factors like geography and transmitter location, some stations cover certain areas of the market better than others. If the county that is moved is one in which a particular station has a clear technical advantage, it could alter the relative standing of stations in the ratings.

Figure 6.3 is typical of one of the first pages you will encounter in a "local television market report." This one happens to come from an Arbitron report on Memphis. Local market reports are the primary vehicle for reporting station ratings. They are "the books" that cause so much anxiety among station personnel. This page contains a good deal of information about the market.

The first thing to notice is a map of the market area. Although ADIs and DMAs divide markets into nonoverlapping areas, other geographic distinctions are also made. Both Arbitron and Nielsen identify a smaller area within the market called the *metro* area. This is the core retail area of the market, and generally corresponds to the Metropolitan Statistical Areas (MSAs) used by the federal government. The map also depicts a much larger area that encompasses the ADI and several counties outside the ADI. This area, which Arbitron calls the Total Survey Area (TSA), includes counties that belong to adjacent markets, but that nevertheless have viewers who watch local market stations.

Below the map is a section called "Estimates of Households in Market." Look under the column labeled "ADI." This reports the households with television as well as estimates of the percent subscribing to cable and owning a VCR. The section below that is called "Television Stations." It reports the stations that are significantly viewed in the market, including

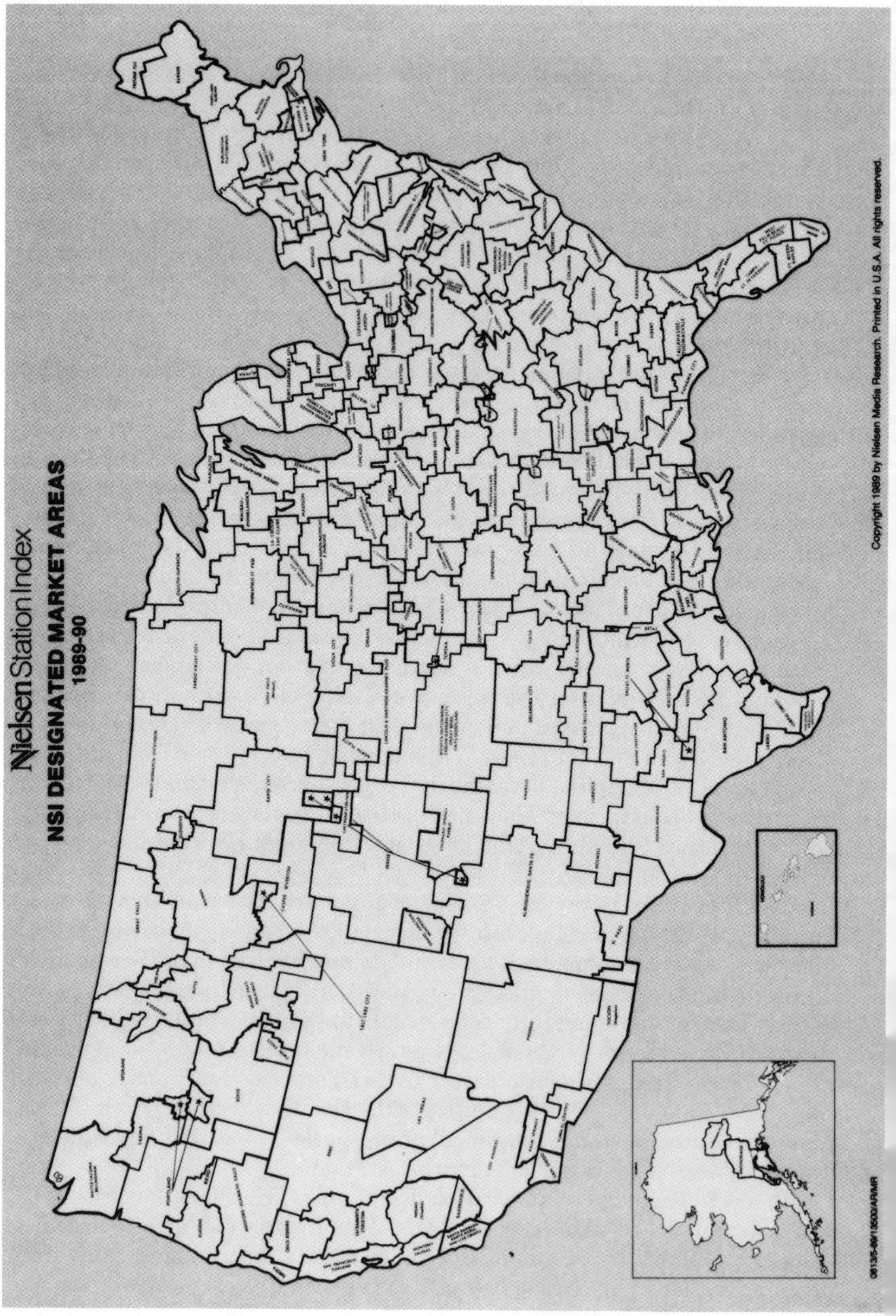
Nielsen Station Index
NSI DESIGNATED MARKET AREAS
1989-90
Copyright 1989 by Nielsen Media Research. Printed in U.S.A. All rights reserved.

RATINGS RESEARCH PRODUCTS

Audience Estimates in the Arbitron Market of

Memphis

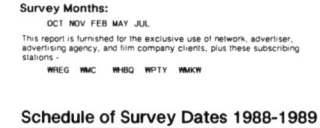

Survey Period: FEB 01,1989 - FEB 28,1989

Survey Months:
OCT NOV FEB MAY JUL

This report is furnished for the exclusive use of network, advertiser, advertising agency, and film company clients, plus these subscribing stations -
WREG WMC WHBQ WPTY WMKW

Schedule of Survey Dates 1988-1989

October	September 28 - October 25, 1988
November	November 2 - November 29, 1988
January	January 4 - January 31, 1989
February	February 1 - February 28, 1989
March	March 1 - March 28, 1989
May	April 26 - May 23, 1989
July	July 5 - August 1, 1989

☐ Metro ☐ ADI ☐ TSA

© Arbitron Ratings Company

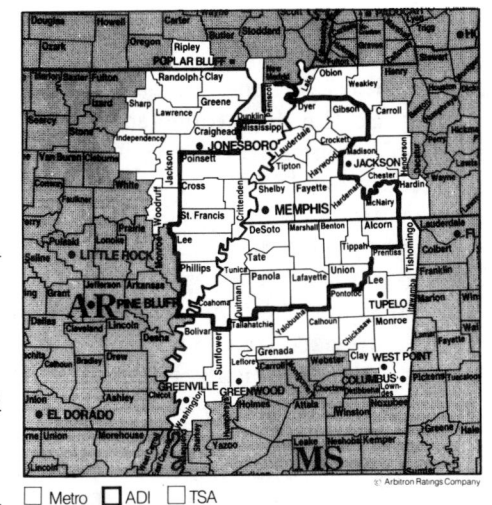

Estimates of Households in Market

	TSA	Pct TVHH	ADI	Pct TVHH	Metro Rating Area	Pct TVHH
TOTAL HOUSEHOLDS	984,600		594,100		352,900	
TV HOUSEHOLDS	970,100	100	586,500	100	349,500	100
MULTISET TV HH	564,200	58	417,200	71	275,500	79
CABLE SUBSCRIBERS	530,400	55	298,200	51	187,800	54
VCR HOUSEHOLDS	558,100	58	356,400	61	222,800	64

Television Stations

Call Letters	Channel Number	Affiliation	Home Home Non-ADI Outside	City of Identification Authorized by FCC
WREG-TV	3	CBS	H	MEMPHIS, TN
WMC-TV	5	NBC	H	MEMPHIS, TN
WHBQ-TV	13	ABC	H	MEMPHIS, TN
WPTY-TV	24	IND	H	MEMPHIS, TN
WMKW	30	IND	H	MEMPHIS, TN
WKNO-TV	10	PTV	H	MEMPHIS, TN
WMAV-TV	18	PTV	H	OXFORD, MS
WTBS	17	TBS	O	ATLANTA, GA

FEBRUARY 1989   INT-1   MEMPHIS

FIG. 6.3. Arbitron market area (reprinted with permission of the Arbitron Company).

super stations and networks received over cable. This page also includes information on the "Schedule of Survey Dates," just to the left of the map. In all but the largest markets, which we will discuss later, ratings data are gathered only during certain times of the year. These occasions are referred to as ratings *sweeps*. A sweep is four weeks long, or roughly a month in duration. All television markets are swept at least four times a year, in November, February, May, and July. Some markets are swept more often. The Memphis ratings book we are using is from the February 1989 sweep.

During a ratings sweep, both Arbitron and Nielsen are placing and retrieving television diaries in households throughout the market. You may recall, however, the standard diary only records one week's worth of viewing. So the data collected in a sweep is actually based on four independent samples drawn in consecutive weeks. In most instances, these data are combined to provide a single monthly estimate of audience size and composition.

A ratings sweep is more that just an occasion for collecting data, however. The dates of each sweep are known well in advance, so local stations can and do adapt their programming to attract the largest audiences possible during each sweep. In television, this may manifest itself in the local news airing particularly sensational stories. Even the networks, which are continuously measured, try to help out their affiliates by running blockbuster movies and heavily promoted mini-series. Sometimes, though, a station will run amok and cross the rather ill-defined line between reasonable promotional efforts, and illegal practices known as hyping or *hypoing*. Indeed, such abuses were one of the concerns motivating the Congressional investigations in the 1960s.

Hypoing can involve any one of a number of activities designed to distort or bias ratings results. For example a station might try to enhance its ratings by directly addressing diary-keepers in its programming, by conducting a survey to learn the identity of actual diary-keepers, or by conducting particularly heavy-handed contests and promotions. If the ratings companies learn of such "special station activities," they may take several different actions, from placing a special notice in the rating book, to deleting the station's audience estimates altogether.

Assuming that data collection goes according to plan, each ratings company will have about 100,000 diaries to process at the end of a nationwide sweep. Sample sizes vary widely from market to market. The largest markets like New York will have household samples of about 1,700. The smallest markets, by contrast, will have just over 200 households in-tab. Response rates also vary from market to market, but average around 45%. In any particular market, information on sample placement and response is contained in the local market report.

Local television ratings are reported in various ways. Television market reports provide audience estimates by daypart, or more discrete time periods, they provide audience trend information for different demographic groups, and they describe audiences for specific television programs. Figure 6.4 is a page on "Time Period Estimates" from the Memphis market.

Across the top are column headings that describe the contents of the numbers directly below. Down the left hand side of the page is information on specific stations and the programs they were broadcasting. It is orga-

Time Period Estimates

DAY AND TIME / STATION PROGRAM	ADI TV HH RATINGS BY WEEKS 1 FEB 01	2 FEB 08	3 FEB 15	4 FEB 22	ADI TV HH RTG	ADI TV HH SHR	ADI TV HH SHARE/HUT TRENDS NOV 88	JUL 88	MAY 88	FEB 88	METRO TV HH RTG	METRO TV HH SHR	ADI RATINGS PERSONS 2+	18+	12-24	12-34	18-34	18-49	21-49	25-54	35+	35-64	50+	WOMEN 18+	12-24	18-34	18-49	21-49	25-49	25-54	W W	MEN 18+	18-34	18-49	21-49	25-49	25-54
	1	2	3	4	5	6	59	60	61	62	8	9	11	12	13	14	15	16	17	18	19	20	21	22	23	24	25	26	27	28	29	30	31	32	33	34	35
▲ RELATIVE STD-ERR THRESHOLDS (1σ) 25%	8	8	8	8	2						4		1	1	6	3	4	2	2	2	1	2	3	2	11	7	3	3	4	4	6	2	8	4	5	5	4
50%	2	2	2	2	-						1		-	-	1	-	1	-	-	-	-	-	-	-	3	1	1	1	1	1	1	-	2	1	1	1	1
THURSDAY																																					
4:30P- 5:00P																																					
WREG THE JEFERSNS	11	11	12	9	11	23	23	28	22	24	12	26	9	8	14	11	9	9	8	7	8	7	7	9	16	10	9	8	8	7	6	7	8	9	7	8	7
WMC JEOPARDY	10	19	17	20	17	36	38	32	29	30	18	39	10	12	4	4	4	6	6	7	17	11	25	14	4	3	5	6	6	8	4	10	5	6	6	6	7
WHBQ WIN-LSE-DR-S	3	2	7	3	3	7	11	12	16	13	4	8	3	3	4	4	4	3	3	3	2	2	1	4	6	6	5	4	4	4	2	1	2	1	1	1	1
WPTY FINDRS KEEPR	1	3	-	4	2	5	6	7	7	5	3	6	2	1	2	2	1	1	1	1	1	1		1	2	2	1	2	1	1	1	1	1	1	1	1	1
WMKW DUCK TALES	3	4	3	5	4	8	4	8	10	6	5	11	3	1	2	2	1	1	1	1	1	1	1	1	3	1	1	1	1	1	1	1	1	1	1	1	1
WKNO PTV	-	1	1	1	1	2	2	3		3	1	2	1				1							1		1		1	1								
H/P/T	42	47	50	48	47		36	35	43	50	47		33	30	33	28	25	25	23	24	33	26	39	35	39	30	28	27	27	29	20	24	21	21	20	20	20
5:00P- 5:30P																																					
WREG COSBY SHOW-S	19	25	18	18	20	33	31	22	18	18	22	36	18	15	26	21	19	17	16	14	13	12	12	16	28	19	18	17	16	16	13	15	19	17	14	13	13
WMC ACTION NWS 5	22	27	32	27	27	44	46	41	40	44	30	48	16	19	6	7	7	10	11	13	27	20	37	20	5	7	11	11	12	14	12	18	8	10	10	11	12
WHBQ USA TODAY	2	1	4	4	3	5	6	9	12	9	3	4	2	2	1	1	1	2	2	2	3	3	3	3	2	3	3	2	3	3	2	1		1	1	1	2
WPTY KNIGHT RDR-S	2	2	1	2	2	2	4	8	7	5	1	2	1	1	1	1	1	1	1	1	1	1		1	1	1	1	1	1	1	1	1	1	1	1	1	1
WMKW SILVR SPNS-S	2	2	1	4	2	4	2	3	11	8	4	6	2	1	3	2	1	1	1	1	1	1		1	2	1	1	1	1	1	1	1	2	2	2	1	1
WKNO PTV	-	1	1	-	1	1				2	1	1	1				1							1		1		1	1								
H/P/T	58	63	66	59	62		50	45	50	61	62		44	44	44	39	37	38	37	37	49	42	57	47	45	38	40	40	39	40	34	42	36	36	34	32	34
5:30P- 6:00P																																					
WREG CBS EVE NEWS	10	13	9	11	11	18	24	19	18	19	13	21	7	8	6	6	6	6	6	6	8	7	10	9	7	7	7	7	7	8	7	6	5	5	5	4	5
WMC ACTION NWS 5	21	27	30	29	27	45	43	41	37	43	29	47	16	19	7	7	8	10	11	13	26	19	36	19	5	6	10	10	11	13	11	19	10	11	11	11	12
WHBQ ABC WRLD NWS	1	1	2	5	2	4	7	10	10	10	2	4	1	1	1	1		1	1	2	2	2	3	2	1		1	1	1	1	1	1		1	1	1	2
WPTY FAMLY TIES 1	6	6	5	6	6	10	9	10	8	10	7	12	5	4	8	6	5	5	5	4	3	4	2	4	9	6	5	5	4	3	4	4	5	6	5	5	5
WMKW FACTS LIFE-S	3	3	4	5	4	6	4	5	13	8	5	8	2	2	5	3	2	2	2	2	1	1	1	2	7	4	3	3	3	2	3	1	1	1	1	1	1
WKNO PTV	-	1	1	-						1	1	1					1							1		1			1								
H/P/T	55	59	63	60	60		50	45	50	65	61		38	41	35	31	32	33	33	33	47	40	57	43	36	31	33	33	32	34	31	39	32	32	32	29	31
6:00P- 6:30P																																					
WREG NEWS 3 6P	11	15	19	11	14	22	26	26	22	20	16	27	9	10	6	8	9	9	9	10	11	10	13	12	7	10	10	11	12	12	11	9	7	7	7	7	7
WMC NBC NGHT NWS	16	20	20	21	19	30	32	33	30	33	19	31	12	14	4	4	5	7	7	8	20	14	29	15	4	5	6	6	7	8	10	14	5	7	8	8	9
WHBQ 6PM EYWT NW	4	4	5	7	5	8	9	9	11	9	6	10	3	3	3	2	2	2	2	3	4	5	5	4	5	2	4	3	3	3	3	2	1	1	2	2	2
WPTY FAMLY TIES 2	8	7	5	5	6	10	8	5	7	7	8	13	5	3	8	6	4	4	4	3	2	3	1	3	11	5	4	4	3	3	3	3	4	4	4	4	4
WMKW A-TEAM	3	5	4	6	4	7	6	3	6	6	5	9	3	3	4	3	3	3	3	3	2	3	2	3	3	3	3	3	3	3	2	2	3	3	3	3	3
WKNO PTV	1	2	-	-	1	1	2	2	1	2	1	2		1		1	1	1	1	1			1	1		1	1	1	1	1							
H/P/T	57	64	70	62	63		57	50	54	68	62		41	45	34	35	36	36	37	38	51	45	61	47	36	36	37	38	38	40	39	43	36	35	36	34	36
6:30P- 7:00P																																					
WREG ENTRTNMT TON	10	13	16	10	12	19	20	20	19	13	16	25	7	8	4	6	7	8	8	9	9	9	9	9	2	7	9	10	11	11	10	7	8	7	6	6	7
WMC WHEEL OF FOR	18	22	25	24	22	35	38	38	37	38	24	38	15	18	9	8	8	9	9	11	23	17	34	19	11	10	10	10	10	12	14	15	7	8	9	8	9
WHBQ FAMLY FEUD-S	6	4	4	7	5	8	12	12	13	17	5	8	4	4	6	4	4	4	4	4	4	5	3	4	8	4	5	5	4	4	3	3	3	3	3	3	3
WPTY NGHT COURT-S	7	5	3	4	5	7	5	4	6	5	5	8	3	3	6	5	4	4	4	3	2	3	1	2	7	3	3	4	3	2	2	3	5	5	5	4	4
WMKW A-TEAM	4	5	4	6	5	7	6	3	6	5	6	9	4	3	4	4	3	3	3	3	3	3	2	3	4	3	3	4	3	3	2	3	3	3	3	3	3
WKNO PTV	-	3	1	1	1	2	3	3	3	5	1	2	1	1		1	1	1	1	1	1	1	1	1		1	1	1	1	1	1	1	1	1	1	1	1
H/P/T	60	62	71	66	65		59	51	55	70	64		47	49	39	40	42	42	43	44	53	49	61	51	39	43	43	45	45	47	45	46	41	41	41	40	41
7:00P- 7:30P																																					
WREG 48 HOURS	8	14	10	7	10	13	11	12	11	9	9	13	6	7	3	5	6	6	6	7	8	7	11	7	2	5	5	6	6	7	6	7	7	6	5	6	7
WMC BILL COSBY	43	42	46	43	44	61	63	57	65	62	46	64	37	34	45	38	34	36	35	34	33	35	29	39	47	40	41	40	40	39	37	28	28	31	30	28	27
WHBQ FINE ROMANCE	4	2	4	4	3	5	6	5	4	10	4	6	2	2	2	2	2	2	2	2	2	2	3	2	3	2	2	2	3	3	3	2	1	1	2	2	2
WPTY SIMON-SIMN-S	1	3	2	1	2	3	2	3	3	2	1	1	1	1	1	2	2	2	2	2	1	1	1	1	2	2	1	1	1	1	2	2	3	2	2	2	2
WMKW M DOWNEY JR	1	2	2	2	2	2	3	2	1	1	3	4	1	1		1	1	1	1	1	2	1	2	1		1	1	1	1	1	1	1	1	1	1	2	1
WKNO PTV	-	2	2	2	2	2	4	4	4	5	1	2	1	1	1	1	1	1	1	1	2	1	2	2	1	1	1	1	1	1		1	1	1	1	1	1
H/P/T	69	72	76	69	71		66	56	64	77	72		57	56	57	55	53	54	54	55	57	57	59	60	62	58	58	59	59	60	58	51	48	49	49	49	50
7:30P- 8:00P																																					
WREG 48 HOURS	8	13	11	7	10	15	10	15	13	11	9	13	6	7	3	5	5	5	5	7	8	8	11	7	2	5	5	5	5	6	7	7	6	6	6	6	7
WMC DIFFRNT WRLD	35	36	41	37	37	56	59	47	59	56	38	57	31	28	40	33	30	30	29	27	26	27	24	32	40	35	34	33	32	32	29	23	26	26	25	23	23
WHBQ FINE ROMANCE	4	2	3	4	3	5	7	5	5	14	4	6	2	2	2	2	2	2	2	2	2	2	2	2	3	3	3	3	3	3	3	1	1	1	1	1	1
WPTY SIMON-SIMN-S	1	2	2	1	2	2	1	3	3	2	1	1	1	1	1	2	2	2	2	1	1	1	1	1	2	2	1	1	1	1	2	2	3	2	2	2	2
WMKW M DOWNEY JR	1	2	2	3	2	3	3	3	1	2	3	4	1	1		1	1	1	1	1	2	1	2	1		1	1	1	1	1	1	1	1	1	1	2	1
WKNO PTV	3	2	1	3	2	3	6	8	5	5	3	4	2	2	1	1	1	1	1	2	2	2	2	2	1	1	1	1	1	1	1	1	1	1	1	2	2
H/P/T	63	67	71	66	67		63	52	60	73	67		52	51	52	50	49	49	49	50	52	51	54	54	54	52	52	52	52	54	53	47	45	46	45	45	47
8:00P- 8:30P																																					
WREG PARADISE	9		13	12	11	17					8	13	7	8	7	6	6	6	6	7	8	8	10	7	7	4	5	6	6	7	7	8	8	7	6	6	7
PRS ADRS-CBS		18			18	27					19	28	9	12	1	3	3	4	5	9	17	13	26	15			3	4		8		8		5	6		10
--4 WK AVG--					13	20	14	20	21	27	11	17	8	9	5	5	5	6	6	7	11	10	14	9	5	4	5	5	6	7	7	8	7	7	6	6	7
WMC CHEERS	20		27	17	21	32					24	37	14	14	18	16	16	17	17	16	13	14	9	14	15	14	16	16	16	15	13	15	17	17	17	16	16
PRES ADDRESS		17			17	26					18	27	10	11	3	7	7	9	10	12	14	11	16	14			12	13		16		8		5	6		7
--4 WK AVG--					20	31	37	32	41	34	22	34	13	13	14	14	14	15	15	15	13	14	11	14	12	14	15	15	16	15	12	13	14	14	14	14	14
WHBQ DYNASTY	13		13	17	14	22					18	27	10	11	10	9	10	10	9	9	11	11	12	14	13	14	13	12	12	12	12	7	5	6	7	6	7
PRS ADRS-ABC		7			7	10					8	12	5	5	9	6	4	4	3	3	5	4	6	7			7	5		5		2		1	1		1
--4 WK AVG--					12	19	20	7	10	15	15	24	8	9	10	9	8	8	8	8	10	9	11	12	14	13	12	10	10	10	10	6	4	5	5	5	5
WPTY GOLD MOVIE	1	7	5	2	4	6	7	11	6	5	4	5	3	3	3	4	4	3	3	3	2	2	1	2	4	5	3	3	3	3	2	3	4	3	3	4	3
WMKW PRIME MOVIE	2	4	2	4	3	4	2	5	1	4	4	6	2	2	3	2	2	2	2	1	2	2	3	2	3	2	2	1	1	1		2	3	2	3	2	2
WKNO PTV	1	2	2	1	2	2	3	5	1	3	2	3	1	1		1	1	1	1	1	2	1	3	2		1	1	1	1	1	1	1	1	1	1	1	1
H/P/T	60	66	72	64	65		65	54	62	73	65		45	48	43	44	45	45	45	47	50	49	54	50	46	49	48	47	48	49	46	45	41	42	43	42	45
8:30P- 9:00P																																					
WREG PARADISE	8		13	11	11	17					8	13	7	7	7	7	7	6	6	7	8	8	9	7	8	5	6	6	6	7	6	8	8	7	6	6	6
PRS ADRS-CBS		17			17	27					18	26	9	12		2	3	5	5	9	17	13	25	15			4	5		9		8		6	6		10
--4 WK AVG--					12	20	15	19	23	28	11	17	7	8	5	5	6	6	6	7	10	9	13	9	6	5	5	6	6	7	7	8	7	7	6	6	7
WMC LETTRMN ANIV	18				18	32					21	39	12	13	20	17	19	18	18	14	10	13	4	11			14	13		11		17		23	22		18
PRES ADDRESS		15			15	23					17	25	8	9	3	6	6	7	8	10	11	9	12	12			11	12		14		6		4	4		5
DEAR JOHN			22	15	18	28					21	31	11	12	12	11	10	11	12	11	12	12	12	13	14	11	13	13	13	13	11	10	10	10	9	9	9
--4 WK AVG--					17	28	30	31	36	33	20	31	10	11	12	11	11	12	12	12	11	12	10	12	11	11	13	13	13	13	10	11	12	11	11	11	10
WHBQ DYNASTY	13		13	17	14	23					18	29	10	11	11	10	10	10	10	10	12	11	12	14	14	15	14	13	13	12	12	8	6	7	7	7	8
PRS ADRS-ABC		8			8	12					10	14	5	5	8	7	6	5	4	4	5	4	6	8			8	7		6		2		1	1		1
--4 WK AVG--					13	20	21	7	11	16	16	25	9	10	10	9	9	9	9	8	10	10	11	12	15	14	12	11	11	11	11	6	5	5	6	6	6
WPTY GOLD MOVIE	1	7	5	2	4	6	8	10	6	5	4	6	3	3	3	4	5	4	4	4	2	2	1	3	5	6	4	4	3	3	2	3	3	4	4	4	4
WMKW PRIME MOVIE	2	3	2	4	3	4	2	5	1	4	4	6	2	2	2	2	2	2	2	1	2	2	3	2	3	2	2	1	1	1		2	2	2	2	2	2
WKNO PTV	1	2	2	1	2	3	3	5	1	3	2	3	1	1		1	1	1	1	1	2	1	3	2		1	1	1	1	1	1	1	1	1	1	1	1
H/P/T	57	64	69	61	63		60	54	63	71	64		42	46	40	42	44	44	44	46	48	47	52	50	46	50	47	47	47	48	43	43	38	40	40	41	43
	1	2	3	4	5	6	59	60	61	62	8	9	11	12	13	14	15	16	17	18	19	20	21	22	23	24	25	26	27	28	29	30	31	32	33	34	35

MEMPHIS 47 THURSDAY FEBRUARY 1989 TIME PERIOD AVERAGES

- * SAMPLE BELOW MINIMUM FOR WEEKLY REPORTING
- ** SHARE/HUT TRENDS NOT AVAILABLE
- \- DID NOT ACHIEVE A REPORTABLE WEEKLY RATING
- ‡ TECHNICAL DIFFICULTY
- \+ COMBINED PARENT/SATELLITE
- ▲ SEE TABLE ON PAGE v

FIG. 6.4. Arbitron time-period estimates (reprinted with permission of the Arbitron Company).

nized by day of the week, and within that, by half hour time periods. This particular page reports viewing on Thursday, from 4:30 p.m. to 9:00 p.m. At the bottom of each half hour time period is a line labeled "H/P/T" which gives the HUT level, PUT level, or the total number of people in the audience in that half hour.

The first four columns of numbers give audience estimates for each of the four weeks that actually comprise a sweep. This can be useful, because sometimes there are programming changes in a sweep that can affect the average rating. For example, in the 8:00–8:30 p.m. time slot near the bottom of the page, a Presidential address preempted regular programming in the second week of the sweep.

The remaining columns report average program ratings and shares of one sort or another. For instance, the fifth and sixth columns contain rating and share information for TVHH in the ADI. If you look in the 7:00–7:30 p.m. time period, you'll see that station WMC aired "Bill Cosby," and achieved a 44 rating and a 61 share. The following page in this ratings book (not shown) contains projected audience estimates, reported in thousands, for the same programs.

We should also point out that at the top of the page, right under the column headings, are two rows of numbers labeled "Relative Standard Error Thresholds." These should serve to remind the users that the numbers reported below are only estimates based on samples, and are therefore subject to sampling error. More specifically, for each column, they indicate the point at which one standard error will constitute either 25% or 50% of an estimate. As you would expect, column estimates based on smaller sample sizes (e.g., women 12–24 vs. women 18+) are subject to more error, hence thresholds are relatively high.

Both ratings companies will make local market reports available in machine readable form. Usually, market reports are stored on computer disks, and read by using a desktop computer. Not only can one read the book this way, but more importantly, the audience estimates contained within can be more easily manipulated. Both ratings companies now market their own software for sorting through what is essentially an electronic version of the market report. The Arbitron software is called "TV Maximizer." NSI sells three packages for manipulating local data called "Spotbuyer," "Postbuy Reporter," and "Audience Analyst." Independent vendors also sell software for the analysis of market report data. The exact capabilities of each package differ, but they can typically locate the relative strength and weaknesses of each station in the market by ranking on various criteria, identify a package of avails to match an advertiser's request, help manage audience inventories, and project audiences based on historical data. Many of these specific analytical techniques are discussed in the last chapter.

In the top 15 to 25 markets, Arbitron and/or Nielsen maintain continuous panels of metered household in addition to their diary-based research. At this writing, the meters in use are passive household meters, as opposed to peoplemeters. There are, however, plans to introduce peoplemeters to major markets. Should that happen, it would presumably eliminate the need for diary keeping in those markets, although the overlaping TSAs of metered and nonmetered markets are problematic. At present, there are between 300 to 500 metered households per market.

Having metered data available affects both the ratings and how they are used. First, as we described in the previous section, meter-based data are employed to adjust the audience estimates derived from diary data. Second, because data collection is fast, meters make it possible to deliver overnight ratings. These are only household level data, but they are reported to one more decimal place than published ratings, and can allow programmers to respond quickly to audience trends. For example, by monitoring overnight during a sweeps period, a station may be able to identify a program that is on the verge of moving to the next whole number in the ratings. Dollar for dollar, going from a 9 to a 10 could be very important. If such a shift seems possible, it might be worth running a few extra promotions in an effort to lock in the higher rating for the next local market report. As was the case with local market reports, both Arbitron and Nielsen sell PC-based software to manipulate electronically delivered overnight data.

Local radio audience estimates are also an important product of the ratings services. Radio ratings, however, are not directly analogous to television ratings. There are differences in how markets are defined and how often they are measured. There are also important differences in the research methods used by the two principal suppliers of local radio ratings. Arbitron bases its ratings on diary data, whereas Birch uses telephone recall techniques. These methodological differences are associated with systematic differences in the audience estimates.

Both services identify more markets than are found in television ratings. Arbitron reports audience estimates for roughly 260 markets. Birch now reports on about 250. All radio ratings books estimate audiences for a metro rating area, which generally corresponds to a governmentally designated metropolitan area. Arbitron routinely reports Total Survey Area (TSA) estimates for the larger geographic area in which radio listening may nonetheless occur. In the top 50 radio market, Arbitron will also report listening in the ADI, as defined by patterns of television viewing. This promotes comparison of radio and television coverage. Obviously, however, with more radio than television markets, this one-to-one correspondence cannot extend beyond the top markets.

Arbitron measures all markets in the spring. Its survey period lasts for

12 weeks, instead of the 4 weeks used for the TV ratings sweep. All estimates in the market report represent an average week in that period. Many markets are measured again in the fall, and the largest 75 to 80 markets are measured four times a year (i.e., spring, summer, fall, winter). With each measurement period extending for 12 weeks, that means that the largest markets are essentially under continuous measurement. Birch's schedule of measurement and report production is even more involved. Basically, the largest 100 markets are continuously measured and supplied with both monthly and quarterly audience summaries, whereas smaller markets are studied less intensively.

More interesting, and more controversial, are the systematic differences that seem to emerge when listening behavior is measured by telephone as opposed to diaries. It has been frequently observed that certain radio formats, like contemporary hit radio (CHR) and urban contemporary, do better in the ratings when those numbers are based on telephone interviews. Other formats, like easy listening, seem to benefit from diary-based measures. Similarly, AM radio's share of the market is higher when estimates are based on diary data.

The best single explanation for these differences appears to be response rates. Overall, Birch reports a response rate of about 60%, whereas Arbitron's rate is roughly 45%. Moreover, nonresponse among diary keepers is especially acute in younger age groups. These listeners tend to favor formats like CHR. This creates a bias in the data that we earlier referred to as *nonresponse error*. The under-representation of younger listeners can, in part, be corrected by weighting the data in favor of younger listeners who do return a diary. Unfortunately, those who respond may have different format preferences than their peers who did not respond, and weighting the data by demographic categories cannot compensate for that sort of difference. Both services report unweighted in-tab sample demographics in comparison with population estimates, and ratings users should be alert to which groups are under- or over-represented.

Those differences aside, the kinds of radio audience summaries available in Arbitron and Birch market reports are similar. Market reports from either supplier feature a large section called either "Target Audience" (Arbitron) or "Target Demographics" (Birch). Here, the radio audience is broken out into 20 or more demographically defined subsets. Within each demographic category, individual station audiences are reported across several dayparts. Figure 6.5 is an example of one such page for the spring Arbitron market report for Memphis.

This page includes station audience estimates among men 18–49 years old. The column headings across the top identify five different dayparts. Other daypart estimates are reported in the pages that follow.

Target Audience
MEN 18-49

	MONDAY-FRIDAY 6AM-10AM				MONDAY-FRIDAY 10AM-3PM				MONDAY-FRIDAY 3PM-7PM				MONDAY-FRIDAY 7PM-MID				WEEKEND 10AM-7PM			
	AQH (00)	CUME (00)	AQH RTG	AQH SHR	AQH (00)	CUME (00)	AQH RTG	AQH SHR	AQH (00)	CUME (00)	AQH RTG	AQH SHR	AQH (00)	CUME (00)	AQH RTG	AQH SHR	AQH (00)	CUME (00)	AQH RTG	AQH SHR
KFTH																				
METRO	2	28	.1	.3					2	40	.1	.3		6			2	17	.1	.4
TSA	2	28							2	40				6			2	17		
KHLS																				
METRO	1	7		.2	1	7		.2	1	15		.2								
TSA	3	29			4	40			1	15			1	11						
KMPZ																				
METRO	23	205	.9	3.5	18	166	.7	3.1	22	205	.8	3.8	7	109	.3	2.3	22	170	.8	4.6
TSA	25	265			24	207			35	303			7	129			24	185		
KRNB																				
METRO	25	279	1.0	3.8	30	250	1.1	5.2	41	233	1.6	7.1	27	246	1.0	8.7	40	236	1.5	8.3
TSA	26	317			32	308			45	273			33	304			40	236		
KSUD																				
METRO		7			2	13	.1	.3	3	6	.1	.5		8			1	6		.2
TSA		7			2	13			3	6				8			1	6		
KWAM																				
METRO	5	40	.2	.8	3	40	.1	.5	6	40	.2	1.0	3	44	.1	1.0	1	7		.2
TSA	13	53			13	53			11	53			3	44			3	20		
WCRV																				
METRO	8	57	.3	1.2		32			1	31		.2	2	10	.1	.6	1	14		.2
TSA	13	72			3	47			3	46			2	10			2	22		
WDIA																				
METRO	51	225	1.9	7.7	32	135	1.2	5.6	42	201	1.6	7.2	28	151	1.1	9.1	40	227	1.5	8.3
TSA	61	250			53	189			52	235			28	167			60	320		
WEGR																				
METRO	124	456	4.7	18.8	125	386	4.8	21.7	96	416	3.7	16.6	26	264	1.0	8.4	73	367	2.8	15.1
TSA	189	645			131	469			127	561			41	365			89	471		
WEZI-FM																				
METRO	18	48	.7	2.7	17	63	.6	3.0	4	40	.2	.7	6	34	.2	1.9	7	51	.3	1.4
TSA	18	48			17	63			4	40			6	34			7	51		
WGKX																				
METRO	88	401	3.4	13.4	67	314	2.6	11.6	64	399	2.4	11.0	23	194	.9	7.4	48	302	1.8	9.9
TSA	95	530			86	418			87	530			30	241			51	342		
WHBQ																				
METRO	10	49	.4	1.5	6	20	.2	1.0	7	52	.3	1.2	3	33	.1	1.0	2	38	.1	.4
TSA	10	49			6	20			7	52			3	33			5	58		
WHRK																				
METRO	66	434	2.5	10.0	49	378	1.9	8.5	67	457	2.6	11.6	73	454	2.8	23.6	81	401	3.1	16.8
TSA	109	643			57	452			95	603			91	567			101	494		
WLOK																				
METRO	32	146	1.2	4.9	23	96	.9	4.0	23	58	.9	4.0	15	48	.6	4.9	15	104	.6	3.1
TSA	32	146			23	96			23	58			15	48			15	104		
WMC																				
METRO	11	116	.4	1.7	5	59	.2	.9	10	119	.4	1.7	3	36	.1	1.0	1	24		.2
TSA	23	134			13	94			13	137			3	54			2	41		
WMC-FM																				
METRO	53	321	2.0	8.0	49	300	1.9	8.5	39	283	1.5	6.7	19	190	.7	6.1	30	279	1.1	6.2
TSA	66	399			64	385			60	384			34	244			37	338		
WREC																				
METRO	6	51	.2	.9	4	34	.2	.7	13	75	.5	2.2	1	39		.3		34		
TSA	7	59			5	41			13	83			1	39				34		
WRVR																				
METRO	28	134	1.1	4.2	31	115	1.2	5.4	26	122	1.0	4.5	4	27	.2	1.3	11	83	.4	2.3
TSA	29	141			34	155			26	122			4	27			12	104		
WRVR-FM																				
METRO	62	306	2.4	9.4	67	297	2.6	11.6	56	317	2.1	9.7	35	235	1.3	11.3	65	317	2.5	13.5
TSA	73	424			72	399			64	401			36	301			70	365		
WXSS																				
METRO	3	36	.1	.5	5	52	.2	.9	5	51	.2	.9	1	19		.3		16		
TSA	3	36			5	52			5	51			1	19				16		
METRO TOTALS	659	2233	25.2		576	1901	22.0		580	2166	22.2		309	1623	11.8		483	1922	18.4	

Footnote Symbols: * Audience estimates adjusted for actual broadcast schedule. + Station(s) changed call letters since the prior survey - see Page 5B.

MEMPHIS — ARBITRON 45 — SPRING 1989

FIG. 6.5. Arbitron target audience (reprinted with permission of the Arbitron Company).

Underneath each daypart heading are four different audience estimates: (a) the projected audience size in an average quarter hour (AQH); (b) the cumulative audience in that same daypart; (c) AQH audience expressed as a rating; and (d) AQH audience expressed as a share. The first two numbers are always reported in hundreds, with the last two zeros understood. Every listener in an AQH will be included in the associated cume audience, but the reverse is almost never true. Therefore, for any given daypart, the cume will always be equal to or greater than the AQH audience.

Down the left-hand side of the table are the stations reported in the ratings book. Usually these are stations assigned to the home market, but if stations assigned to neighboring markets have significant audiences, they will appear below a dotted line on the same page. Each station has an estimated audience in the both the metro and TSA for the market. For example, station WMC-FM has an estimated metro AQH audience of 5,300 and a TSA audience of 6,600 in AM drive time (i.e., 6 a.m.–10 a.m., Monday–Friday). The corresponding cume audience estimates are, of course, much larger.

A Birch market report also has a rather large section called *Rankers,* in which it takes the same sort of audience estimates, and uses it to rank order stations by the size of their AQH or cume audiences. As we discuss in the final chapter, one of the most common analytical techniques is to make a comparison of audience size among competing stations or media vehicles. This section does that, allowing stations to quickly determine if they are "Number 1" in any particular daypart or demographic group.

Both suppliers will report other data, including hour-by-hour listening estimates, trend data, exclusive cumes, cume duplications, and, because much listening occurs outside the home, location of listening estimates. We discuss these audience summaries further in the last chapter. Both suppliers also repackage their market-level data to create products that provide national radio listening information. These provide some basis for assessing overall listening trends, or for comparing the performance of certain types of stations.

Just like television market data, the information contained in radio market reports is now available on computer disks. Arbitron markets the software needed to manipulate the data under the name "Radio FasTraQ," whereas Birch has a system it calls "Birch Plus." Birch also has a system called "Radio Spot Buyer," which is designed primarily for agencies and advertisers, and is similar to the television spot buyer program developed by Nielsen Media Research which was mentioned earlier.

SYNDICATED PROGRAM RATINGS

The third major market for ratings information is in the area of syndicated programming. This market has grown tremendously in recent years. On one hand, barter syndication has opened up many new channels through which national advertisers can reach viewers. It is important, therefore, to estimate the size and composition of these audiences. On the other hand, the growth of independent stations, and the increasing inventory of non-network programming, means that stations are in a position to program more and more of what they air. As a result, programmers must be alert to the ratings performance of the different options available to them.

Both Nielsen and Arbitron serve this marketplace. Both use the data they have collected for other purposes to develop syndicated program ratings. In Nielsen's case, it can draw on both the peoplemeter data it uses for its NTI reports, and the diary-based data employed to generate most of its NSI market reports. Because Arbitron does not have a national ratings panel in place, its syndicated program reports are all based on data gathered for its local market reports.

Each company issues a report on syndicated programs based on its local market data. Nielsen's is called the *Report on Syndicated Programs;* Arbitron's is the *Syndicated Program Analysis*. These appear four times a year, after each of the four major ratings sweeps. Basically, the ratings suppliers extract from their local market data the ratings performance of every syndicated program. These are then organized by program so that the users can see the program's average performance across all markets, as well as how it did in each market that carried the show.

Figure 6.6 is a page from Nielsen's *Report on Syndicated Programs*. It is the first of several pages that describe the audience for the "Oprah Winfrey Show," a popular syndicated talk show. In the upper left-hand corner of the page is information on the program's coverage, distributor, and so forth. This program aired on stations that, taken together, reach 99% of all TVHH in the United States. The upper third of the table summarizes how "Oprah" does across all of those markets. Most stations ran it in the "early fringe" daypart, but a few aired it during "daytime." In either case, it had audience shares that averaged in the high 20s or 30s.

The lower two thirds of the table, and the pages that follow, report audience estimates for "Oprah" in each of the markets that carried the program. In any given market, the audience for a syndicated show is affected by both the competition and the audience of the "lead-in" program. Such programming information is, therefore, provided in the report. That way, if station personnel are trying to evaluate the performance of a

MARKETS REPORTING 197
STATIONS REPORTING 199
TOTAL TV HH'S IN DMA'S 89,397,360
DMA % OF U.S. 99
EPISODES AVAILABLE 220
DIST: KING WORLD PRODUCTIONS
TYPE: INTERVIEW

REPORT ON SYNDICATED PROGRAMS

NSI AVERAGE WEEK ESTIMATES

FEB 1989

OPRAH WINFREY SHOW

60 MIN.

SUMMARY BY DAYPARTS

DAYPART	DMA HOUSEHOLD SHARES BY MARKET RANK 1-25 NO.OF DMA'S	1-25 % SHARE	26-50 NO.OF DMA'S	26-50 % SHARE	51-100 NO.OF DMA'S	51-100 % SHARE	101+ NO.OF DMA'S	101+ % SHARE
DAYTIME(M-F) †	5	27			3	36	9	37
EARLY FRINGE(M-F)	20	28	25	37	46	32	86	36
PRIME ACCESS(M-SAT)								
PRIME(S-S)								

DAYPART	DMA HOUSEHOLD SHARES BY MARKET RANK 1-25 NO.OF DMA'S	1-25 % SHARE	26-50 NO.OF DMA'S	26-50 % SHARE	51-100 NO.OF DMA'S	51-100 % SHARE	101+ NO.OF DMA'S	101+ % SHARE
POST PRIME(S-S)							1	
WEEKEND DAYTIME(S&S)								
WEEKEND PRE-PRIME(S&S)								
AVG. ALL TELECASTS	25	28	25	37	49	33	94	36

DAYPART	NO. OF MKT's	NO. OF DMA's	% U.S. TV	DMA H/H AVG Q.H. RTG	DMA H/H SHR	TOTAL HSHLDS (000)	WOMEN TOTAL (000)	WOMEN TOTAL V/CVH	WOMEN 18-49 (000)	WOMEN 18-49 V/CVH	WOMEN 25-54 (000)	WOMEN 25-54 V/CVH	MEN TOTAL (000)	MEN TOTAL V/CVH	MEN 18-49 (000)	MEN 18-49 V/CVH	TEENS TOTAL (000)	TEENS TOTAL V/CVH	CHILDREN TOTAL (000)	CHILDREN TOTAL V/CVH
DAYTIME (M-F)†	17	17	16	9	28	1383	1205	87	652	47	604	44	366	26	180	13	63	5	59	4
EARLY FRINGE (M-F)	181	177	83	12	32	10197	8907	87	4715	46	4220	41	3054	30	1473	14	760	7	464	5
PRIME ACCESS (M-SAT)																				
PRIME (S-S)																				
POST PRIME (S-S)	1	1		<		<														
WEEKEND DAYTIME(S&S)																				
WEEKEND PRE-PRIME(S&S)																				
TOTAL DAY	197	193				11575	10108		5365		4822		3419		1652		822		523	
AVG. ALL TELECASTS				12	32	59	51	87	27	46	24	42	17	30	8	14	4	7	3	5

FOUR WEEK AVERAGE TIME PERIOD AUDIENCES (THIS PROGRAM vs. PRECEDING HALF HOUR) — PROGRAM AUDIENCE SECTION (SYNDICATED PROGRAM ONLY) — COMPETING FOUR WEEK AVERAGE TIME PERIOD AUDIENCES

LINE 1: MARKET, T.Z., REPORTABLE STATIONS ON AIR
LINE 2: STATION CH. NET., TOTAL DAY DMA SHARE
LINE 3: DAY, START TIME, NO. OF T/CS.
LINE 4: LEAD-IN-PROGRAM

LINE 1–4	DESIGNATED MARKET AREA DMA % H/H RTG (1)	DMA % SHR (2)	PERSONS SHARE % ‡ WOMEN TOTAL (3)	WOMEN 18-49 (4)	WOMEN 25-54 (5)	MEN TOTAL (6)	MEN 18-49 (7)	MEN 25-54 (8)	TNS TOTAL (9)	CHD TOTAL (10)	DMA AREA % H/H RTG (11)	DMA AREA % SHR (12)	(000) VS V/100VH	STATION TOTALS TOTAL HOUSEHOLD (13)	PERSONS (000) & V/100VH TOTAL ADULTS (14)	WOMEN TOTAL (15)	WOMEN 18-49 (16)	WOMEN 25-54 (17)	MEN TOTAL (18)	MEN 18-49 (19)	TEENS 12-17 (20)	CHILD 2-11 (21)	CORRESPONDING TIME PERIOD—3 HIGHEST COMPETING STATIONS: STATION PROGRAM	DMA AREA % H/H RTG (22)	DMA AREA % SHR (23)
ABILENE-SWTWATR CE 3																									
KTAB CH.32 C 25%																									
M-F 3.00P 20T/C	8	29	35	39	27	26	24	21	1		8	29	(000)	9	11	9	4	3	2	1			KRBC #SNT BRB/INSD D	5	19
#GUIDING LIGHT	2	9	8	3	3	5	12	14					V/CVH		116	93	44	30	23	10			KTXS #BWTCHD /WD WDP	2	9
ADA-ARDMORE CE 2																									
KXII CH.12 CN 31%																									
M-F 3.00P 20T/C	14	37	41	40	39	24	23	26	37	3	14	37	(000)	21	22	18	8	7	4	2	1		KTEN GERALDO	8	21
#GUIDING LIGHT	12	35	39	43	37	22	28	28	44	5			V/CVH		106	85	36	35	21	10	5				
ALBANY-SCH-TROY EA 6																									
WNYT CH.13 N 20%																									
M-F 4.00P 20T/C	9	28	39	39	41	25	20	19	15	2	9	28	(000)	47	56	42	21	19	15	6	4	1	WRGB #BIG MONEY MOV	4	13
SANTA BARBARA	5	19	21	29	28	11	10	9	32	1			V/CVH		121	89	44	41	32	13	8	3	WTEN+#CRRNT F/FCTS F	4	12
																							WXXA DCK TLS/BRD BN	4	10
ALBANY, GA EA 5																									
WALB CH.10 N 49%																									
M-F 4.00P 20T/C	24	54	65	67	63	45	45	43	54	21	24	54	(000)	43	49	38	22	16	11	7	7	4	WTSG DCK TLS/DBL DR	5	11
SANTA BARBARA	23	62	69	71	67	56	60	53	51	41			V/CVH		113	89	52	37	24	16	15	10	WCTV #PPL'S C/JPRD	3	8
																							WVGA BRVSTRR/JTSNS	1	2
ALBUQ-SANTA FE MT 6																									
KOB CH. 4 N 20%																									
M-F 3.00P 20T/C	8	31	36	34	37	22	25	25	19	3	8	31	(000)	32	34	27	16	13	7	4	2	1	KOAT+ INSD DT/JUDGE	4	16
NW DATING GAME	3	14	14	16	17	16	25	17	27	8			V/CVH		104	83	48	40	21	14	6	3	KGGM BST-LV /FML FD	3	14
																							KGSW SCB D /ALVN&C	2	9
ALEXANDRIA, LA CE 4																									
KALB CH. 5 N 42%																									
M-F 9.00A 20T/C	18	61	67	64	61	59	50	58	83	55	18	61	(000)	28	31	23	15	13	8	6	1	2	KLFY #FML FD-/NW CRD	3	10
TODAY SHW	10	42	49	44	43	55	50	44	42	2			V/CVH		110	82	53	46	27	19	3	9	KLAX #GERALDO	2	8
																							KLPA INSTRUCTNL TV	<<	
AMARILLO CE 5																									
KAMR CH. 4 N 22%																									
M-F 3.00P 20T/C	9	31	37	46	46	30	36	38	25	5	9	31	(000)	17	18	14	10	9	4	2	1	1	KVII+#DONAHUE	8	26
SANTA BARBARA	6	24	29	36	31	24	25	26	12	17			V/CVH		106	81	59	53	25	15	4	4	KCIT WD WDPC/ALVN&C	3	9
																							KFDA #JUDGE /SPRR C	3	9
ANCHORAGE YU 6																									
KIMO CH.13 A 17%																									
M-F 3.00P 20T/C	4	21	33	34	24	14	16	17	4	7	4	20	(000)	4.3	4.9	4.1	4.1	2.0	.8	.8	.2	.2	KTUU SANTA BARBARA	5	26
#GENRL HOSPITAL	3	17	24	22	21	6	6	6	5	12			V/CVH		113	95	94	46	19	19	4	4	KTVA VARIOUS	1	8
																							KAKM VARIOUS	1	8
TUE 10.00P 1T/C	1	3	3	3	3	2	1	2			<		(000)	<									KTUU CHNNL 2/CHRS	17	41
#GENRL HOSPITAL	3	17	24	22	21	6	6	6	5	12													KTVA MASH /CRRNT	11	26
																							KTBY ARSENIO HALL	4	9
MARKET AVG.											*4*	*18*	*(000)*	*4.1*	*4.7*	*3.9*	*3.9*	*1.9*	*.8*	*.8*	*.2*	*.2*			
													V/CVH		*113*	*95*	*94*	*46*	*19*	*19*	*4*	*4*			
ANNISTON CE 2																									
WJSU CH.40 C																									
M-F 3.00P 20T/C													(000)	13	15	12	5	4	3	1	1	1	WNAL FN HS /JTSNS		
#GUIDING LIGHT													V/CVH		116	91	38	35	25	11	6	4			

For explanation of symbols, see lead page.

1379

FIG. 6.6. NSI report on syndicated programs (reprinted with permission of Nielsen Station Index).

program in their market, they can look for an appropriate comparison by finding a similar market situation.

Nielsen provides a more elaborate report of program performance on a market-by-market basis in a related service it calls *Cassandra*. Like the *Report on Syndicated Programs, Cassandra* is based on data collected by NSI. However, it sorts and ranks these data in many more ways, often in response to client requests. It is also marketed by a separate division called the Nielsen Syndication Service (NSS). Syndicators, program producers, and large ad agencies are the most frequent users of this report, although stations will occasionally buy it as well. Station rep firms have also developed software to quickly analyze syndicated program performance, because they often advise their clients on program acquisitions, as well as sell available spots to national advertisers.

NSS also markets a pocketpiece that looks very much like the NTI pocketpiece. Like the NTI version, this report is based on the national peoplemeter sample and is issued once a week. Unlike the NTI version, however, it is intended to provide national audience estimates for syndicated programs. The NSS pocketpiece, along with the other reports reviewed here, is used by ad agencies and syndicators to negotiate deals in barter syndication.

CUSTOMIZED RATINGS REPORTS

For the most part, the ratings products we have reviewed so far have been the standardized offerings of the major ratings companies. Usually, they appear as published reports, although increasingly, such reports are delivered in a form that computers can read. In either case, they are reports designed to answer the most common of the research questions we developed in chapter 1. A standard market report or pocketpiece tells you how many people watched a program and who they were—at least in terms of the audience's age and gender. As such, standardized reports are quite useful to most of the people who use ratings data.

As we noted in the first section of the book, however, there are a great many questions that can be addressed with a creative analysis of ratings data. Often, these questions are so specialized that they simply do not justify the publication of a standardized report. Nevertheless, if there are paying customers who want something that cannot be found in a ratings book, ratings firms have ways to accommodate them. Customized ratings reports are created using one of three methods, distinguished by where the data in the report comes from. First, the ratings company can arrange for clients to dip into the company's database and analyze that information in a special way. Second, the ratings data collected in the usual way can be combined with data available from other sources. Third, ratings companies can actually go out and gather more data than they would otherwise collect.

The first option for creating a customized analysis is the most common. Standardized reports only scratch the surface of the analytical possibilities offered by a ratings database. In the last section of the book, we discuss how such analyses can be conceptualized. For now, we concentrate on how and why ratings companies make this option available to clients.

Suppose you were a programmer interested in knowing whether the audience for a syndicated game show stays tuned and watches the local news that follows it. No ratings book published in the United States will give you the answer. Even if the game show and news have exactly the same rating, you cannot tell whether the same people watched both programs. Yet, if you could look at the diaries the ratings company collected, you would be able to figure it out, because the diaries track individuals from one time period to the next. In other words, the ratings company has the information to answer your question; it is just a matter of gaining access to the appropriate data.

Although ratings companies will occasionally sell individual-level data to clients, more often they control clients' access to that kind of data through specially designed computer programs. Typically, users connect to a ratings company computer, and then use the company's software to extract the customized analysis that is needed. This is different from manipulating an electronic ratings book on a desktop computer, because it requires the user, or his or her representative, to go "on-line" and address a larger and more flexible database in some remote location.

Nielsen and Arbitron both make these services available to clients. In essence, they give the user access to all the information contained in the diary data base. Nielsen's NSI Plus deals with local TV ratings. It produces analyses of reach and frequency, audience flow, and offers a myriad of nonstandard demographic and geographic breakouts of the audience. Arbitron's counterpart is called Arbitron Information on Demand (AID). At the national level, Nielsen has an analogous service it calls the Cume Facility. It permits most of the same analyses that the diary-based services provide, but, because its based on peoplemeter data, it can produce cumulative analyses over periods longer than a week.

The array of customized services gets more confusing when new sources of data are introduced into the mix. Recall that advertisers are most often interested in what audience members are likely to buy. For this reason there is considerable pressure on the ratings companies to introduce some sort of produce-usage data into the ratings database. Although the single source technology we described earlier may be the most powerful tool for producing these data, those systems are not fully deployed. More typically, product usage data, along with information on lifestyles, homeownership, and so forth are added to ratings data after the fact.

This is done by matching audience behavior in a very small geographic area, usually a zip code, to other information about that area. Zip codes tend to be relatively homogeneous in composition. For example, some areas are known to be affluent, others poor. Some neighborhoods have large owner-occupied homes, others have a lot of rental units. This information, along with product purchase information is used by companies like Donnelley Marketing Information Services to identify certain "clusters" or categories of zip codes based on their similarities. By assuming that a diary keeper living in a particular kind of area is like others in that area, it is possible to associate ratings data with other variables not in the original database.

Both NSI Plus and AID offer clients access to this zip code-based information. Both Nielsen and Arbitron have also used the availability of this sort of product/lifestyle information to develop services that lie somewhere between an electronic ratings book, and on-line service to a mainframe. Arbitron's version is called Product Target AID. With this system, the user requests certain information about a market area that is then "downloaded" to a personal computer. The PC retains that information, and has the software necessary to manipulate it. Using this system, for example, a station could show a local bank that the station's evening news has a large number of men ages 25–54 who live in affluent areas that are known to invest in financial services. Nielsen offers a similar service called TV Conquest. We discuss these sorts of analyses in greater detail in the final chapter.

The last kind of customized research available from the ratings services involves collecting additional data at the behest of the client. Of course, if the price is right, a ratings company might be persuaded to gather almost any kind of audience data, but two methods of new data collection are worth mentioning here. First, even though it is not their standard method of data collection, both Arbitron and Nielsen will conduct telephone coincidentals. This gives a client the option of getting a ratings report from a major supplier, especially when there is no ratings sweep in progress. Second, it is possible to arrange for diary keepers to be interviewed after their diaries have been collected. By asking questions of a diary keeper, and then matching those responses with the diary record, new insights into the behavior of the audience may be possible. In either case, because new data must be gathered for a single client, these services are not inexpensive.

Although customized ratings reports can provide analysts with many insights that would not otherwise be available, the users of these reports should exercise caution in the interpretation of the numbers they contain.

Remember that the ratings companies are in business to make a profit, and that finding new ways to exploit or resell their existing databases represents a golden opportunity. Remember also that customized reports are by their nature not subject to the same on-going scrutiny of a syndicated report. Ratings companies may very well give a buyer the kind of report asked for, even if it does not make good sense as a piece of research. We have seen, for example, customized market areas constructed from a hand-picked group of counties with too few diaries in-tab to offer reliable audience estimates. In evaluating any ratings report, but especially a customized product, the user must be sure he or she understands the research design upon which the data are based.

BUYING RATINGS DATA

The cost of ratings data varies greatly. A television station in a small market might spent as little as $12,000 a year to get basic ratings reports. An affiliate in a major market might spend close to $1 million on ratings and related services. A broadcast network or large advertising agency will spend much more. There are a number of factors that affect the cost of ratings data, and prices may well be subject to negotiation—especially if there is competition among suppliers.

The most important determinant of price is market size. All things being equal, stations in smaller markets can expect to pay less for ratings than stations in big markets. In part, this is a reflection of the cost of data collection. But a ratings service cannot always price to cover its costs. As we noted in chapter 4, different markets have a history of using different ratings services as the accepted "coin of exchange." Even if it cannot make a profit in a particular market, a service will continue to conduct surveys and price competitively, hoping to make up the difference elsewhere.

Within a given market, there may also be differences in the cost of ratings to different clients. Agencies typically pay less than stations. In fact, in local market research, broadcasters account for about 90% of ratings service revenues. Different stations may also pay different amounts depending on whether they are an independent or an affiliate, a UHF station or a VHF station. Generally, stations with lower circulation receive some sort of discount. Although we have never seen any analysis of this, it is likely the the price stations pay for basic ratings data varies in about the same way that stations base their own rates on these audience estimates. The larger the audience the higher the price.

The length of the contract a client signs can also affect prices. Those who sign long-term contracts should get a discount. A station's subscription to a ratings service will usually run from 3 to 5 years. In metered markets, however, longer commitments may have to be struck, in advance, to induce the ratings company to establish the service.

Academic users can also get special pricing consideration. Nielsen has set packages of both NTI and NSI data designed for educational institutions. Arbitron provides miscellaneous reports to academics upon request. It has also established an archive of its ratings at the University of Georgia. Unfortunately, Nielsen has no public archive of its data, although individual Nielsen offices may maintain informal collections.

Generalizing about the cost of customized ratings reports or access to ratings databases is even more difficult. Despite the analytical possibilities offered by such research, these still account for only a modest portion of ratings service revenues. To learn more about them, or the specific cost of any ratings product, you must deal with the ratings services directly.

Occasionally, a ratings company and one of its clients will have serious differences. A station might be suspected of inappropriate practices during a sweep, or a ratings company might be suspected of mishandling some aspect of the research process. Sometimes a good deal of money can ride in the balance. Although going to court is always a possibility, the parties may find it advisable to opt for a less costly solution. If normal channels of communication fail, the Electronic Media Rating Council (EMRC) can invoke mediation procedures that involve representatives from the appropriate industries and trade associations. The addresses of the ratings services, and the EMRC are provided in Appendix A.

RELATED READINGS

Beville, H. M. (1988). *Audience ratings: Radio, television, cable* (rev. ed.). Hillsdale, NJ: Lawrence Erlbaum Associates.

Fletcher, A. D., & Bower, T. A. (1988). *Fundamentals of advertising research* (3rd ed.) Belmont, CA: Wadsworth.

Poltrack, D. F. (1983). *Television marketing: Network, local, cable.* New York: McGraw-Hill.

PART III

ANALYTICAL TECHNIQUES

A FRAMEWORK FOR ANALYSIS 7

Ratings data come in many different forms and have a wide variety of applications. This abundance may be a bit overwhelming. How does one make sense of all those numbers? What is a high rating, or what is a low one? What is an unusual or important feature of audience behavior, and what is routine? In this chapter we offer a framework for evaluating and analyzing the information contained in ratings data. The emphasis here is on broad concepts and theories. This approach is intended to give readers a sense of perspective on the audience, to help them see "the forest" instead of an endless succession of trees.

Perhaps it is best to begin this exercise by reminding ourselves what ratings data really are. The information collected by the ratings services may be vast in size, and reported in a great many ways, but conceptually it is rather straight forward. The database itself is simply a record of people's reported exposure to electronic media. Developing a framework for analyzing these data, then, requires that we have an understanding of people's media use. If we know what determines exposure to electronic media, if we can predict the patterns of use that are likely to emerge under given circumstances, then we have a way of interpreting the numbers that confront us.

Theories are the tools we use to explain and predict behaviors such as these. A theory is nothing more than a tentative explanation. For many people, the word "theory" seems to imply irrelevance, but as researchers are fond of pointing out, "there's nothing as practical as a good theory." To know if a theory is any good, we must test it. We must determine that it can, in fact, predict or explain what we actually observe in the world around us. Throughout much of this chapter we move back and forth between observations of how audiences behave, and theories or explanations of that behavior. As we do, you are invited to judge for yourself the utility of the theories we encounter.

The chapter is divided into four sections. The first takes a closer look at just what a ratings analyst is trying to assess—exposure to media. The next two sections summarize the major determinants of people's exposure to radio and television—categorized as audience factors and media factors. The last section presents an integrated model of audience behavior to provide a broad framework for evaluating audience information.

EXPOSURE TO MEDIA

Ratings data are a record of people's exposure to electronic media. As we noted in chapter 5, the practice in the industry has been to define exposure as program choice or tuning behavior, rather than as attention or involvement. Taking that as a given, if we study a properly drawn sample of individuals and accurately measure each one, we can have considerable confidence in our ability to describe exposure to radio and television. Of course, the ratings services encounter various problems in sampling, measurement, and data processing. All of these take a toll on the accuracy of the data. But even ratings users who are aware of error in the data, tend to take the numbers at face value in their day-to-day work. For the most part, that is our approach. When substantial methodological problems or biases suggest a qualified interpretation of the data, it is noted, but otherwise, we treat the ratings as valid measures of exposure.

Individual Versus Mass Behavior

Audience analysts are almost always concerned with the behavior of large masses of people. We usually do not care whether Bob Smith sees the early evening newscast, but we do care how many men ages 18–49 will be watching. This interest in mass behavior, which is typical of much social scientific research, is actually a blessing. Trying to explain or predict how any one person behaves, moment to moment, day to day, can be an exercise in frustration. After all, human beings are complex creatures with different moods, impulses, and motivations. Strangely however, when you aggregate individual activities, the behavior of that mass is often quite predictable.

Consider, for example, the birth of a child. If you were asked to predict whether a pregnant woman would give birth to a boy or a girl, your odds of guessing correctly would be about 50/50. On the other hand, if you were asked to predict what percent of babies born in the coming year will be female, you could do so with great accuracy. You need not predict the outcome of each individual case to predict an outcome across the entire population. In the same sense, we do not need to know what every member of a ratings sample will do on a given evening to predict how many households will be using television.

One important consequence of focusing on the mass, rather than individuals, is that audience behavior becomes much more tractable. We can identify stable patterns of audience size and flow. We can develop mathematical equations, or models, that allow us to predict audience behavior. Some have even gone so far as to posit "laws" of viewing behavior. These laws, of course, do not bind each person to a code of conduct. Rather, they are statements that mass behavior is so predictable as to exhibit lawlike tendencies. This kind of reasoning underlies many of the analytical techniques we encounter in the last chapter. (For a more general discussion of mass behavior, see McPhee, 1963.)

Gross Versus Cumulative Measures

We have already encountered many ways to measure or somehow quantify radio and television audiences. Some of these are routinely reported by the ratings services, others are routinely calculated by ratings users. It is useful, at this point, to draw a rather broad distinction between these various audience measurements and indices. We call one group *gross measures,* whereas the other group is labeled *cumulative measures.* The distinction has to do with whether we must track individual behaviors across time. If an audience summary does not depend on tracking, it is a gross measure. If it does, it is cumulative. This temporal quality in the data defines a fundamental distinction that is carried through the rest of the book.

Gross measures of exposure include estimates of audience size and composition made at a single point in time. The best examples are ratings and shares. In effect, these are snapshots of the population that estimate how many people listened to a station in an average quarter hour, or watched a program in an average minute. Projections of total audience size, HUT and PUT levels, belong in this category as well. Gross measures of exposure can also include secondary calculations derived from other gross measurements. Gross ratings points (GRPs) are such calculations. You will recall that GRPs are just a summation of individual ratings over a schedule. Simple cost calculations, like cost per point (CPP) and cost per thousand (CPM) can, similarly, be thought of as a gross measures.

Gross measures are the most common summaries of audience. The majority of numbers reported in a ratings book are of this type. As a result, they are the best known and most widely used of audience measurements. Useful as they are, however, they all fail to capture information about how individual audience members behave over time. That kind of behavior is expressed in cumulative measures.

The most familiar example of the second group of audience measurements is a station's cumulative audience, or cume. To report a weekly cume audience, a ratings company must sort through each person's media use for a week, and summarize the number who used the station at least once. Analogous audience summaries are called circulation, reach, and unduplicated audience. Another cumulative measure that has become increasingly familiar to advertisers is frequency. You will recall that this summarizes how often an individual sees a particular advertising message over some period of time. Measures of audience flow, or program audience duplication, likewise, depend on tracking individual media users over time.

With the exception of the various cume ratings described in the previous chapter, cumulative measures or analyses are less common than gross measurements. These analyses, however, may be useful in a variety of applications. For example, a programmer studying audience flow, or an advertiser tracking the reach and frequency of a media plan is concerned with how the audience is behaving over time. Indeed, as we suggested in chapter 3, this sort of tracking can be illuminating for social scientists interested in any number of questions.

To get a clearer picture of the difference between gross and cumulative measures, and to begin to appreciate the analytical possibilities offered by ratings data, consider Fig. 7.1. The large box in the upper left hand corner represents a simplified ratings database. The data are from a hypothetical sample of 10 households. These are numbered 1 through 10, down the lefthand column. The media use of each household is measured at 10 points in time, running from Time 1 to Time 10 across the top.

In practice, of course, a ratings sample would be much larger, including hundreds or thousands of units of analysis. Those units could be individual people or households, as indicated in the figure. There would also be many more points in time. For example, a standard television diary divides each of 7 days into 80 quarter hours. That means that each person is measured across 560 (i.e., 7 X 80) points in time, rather than the 10 we have illustrated. Now try to imagine how many points in time we could identify in peoplemeter data that track viewing moment to moment over a period of years!

Figure 7.1 portrays television viewing in households, but radio lis-

A FRAMEWORK FOR ANALYSIS

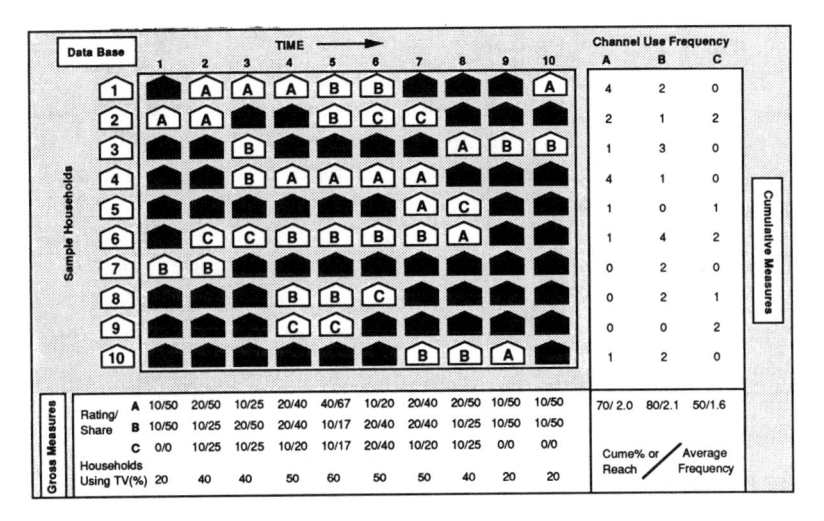

FIG. 7.1. Gross versus cumulative measures in ratings data.

tening data could be conceptualized in much the same way. In our illustration, we have assumed a three-station market, which means that each household can be doing one of four things at each point in time. They can be watching Channel A, Channel B, Channel C, or nothing at all. These behaviors are indicated by the appropriate letters, or a blackened box, respectively.

The most commonly reported gross measures of exposure are shown in the box directly under the database. Each column of data is like Fig. 1.1, and is treated in the same way. Hence, Channel A has a rating of 20 and a share of 40 at Time 4. All one needs to do is look down the appropriate column. Whatever happened before or after that time period is irrelevant to the calculation of a rating (if not the size of the channel's audience).

The box on the right-hand side includes common cumulative measures. To calculate these, we must first examine each household's viewing behavior across time. That means moving across each row in the data base. The first household, for example, watched Channel A four times, Channel B two times, but never watched Channel C. Moving down each channel's column of cumulative viewing, we can then determine its reach, or cume. Each channel's cumulative audience is expressed as a percentage of the total sample who viewed it at least once over the 10 points in time. Therefore, the first household would be included in the cume of A and B, but not C. Further, among those who did view a channel, if we compute the arithmetic average of the numbers in the column, we can report the mean frequency of viewing. This is essentially what an advertiser does when calculating reach and frequency, with the relevant points in time being determined by when a commercial message runs.

Studies of program audience duplication can also be executed from this database. For example, we might be interested in how well Station A retains an audience from one show to the next. We could determine that by seeing how many people who watched Station A at one point in time continued to watch the program that aired after it. For that matter, we could compare any pair of program audiences to assess repeat viewing, audience loyalty, and so on. In each case, however, we would have to track individual households across at least two points in time. Hence, we would be doing a cumulative analysis of exposure.

Depending on the kind of question he or she wants to answer, a ratings user will be involved in interpreting gross measures, cumulative measures, or analyzing numbers that are derived from these two ways of defining exposure. As you will see, there are a large number of analytical techniques that can be organized in this way. To exploit those techniques to their fullest, however, we must develop a better understanding of the factors that shape audiences from moment to moment.

Exposure to the electronic media can be thought of as the interface between the audience and the media. To explain how that interface takes shape and how it changes over time, we need to consider two broad categories of factors. These are, logically enough, audience factors and media factors. Each has a substantial effect on patterns of exposure. Within each category, we have made a further distinction between structural and individual determinants. Although the latter distinction is sometimes hard to make, it is intended to highlight differences in the levels of analysis. It also reflects traditional divisions in research and theory on media exposure. By structural determinants, we mean factors that are common to, or characteristic of, populations. These are "macrolevel" variables typically conceptualized as common to markets or masses of people. Individual determinants are factors descriptive of a person or household. They are "microlevel" variables typically conceptualized as varying from person to person.

AUDIENCE FACTORS

Structural-Level Determinants. The first structural audience factor that shapes exposure to radio and television is the size and location of *potential audiences*. Obviously, no station or network can have an audience larger than the size of the relevant market population. The population, in effect, sets an upper bound on the audience for any program service.

We have noted that each major ratings service divides the country into well over 200 local market areas (see Appendix C). Clearly, the potential audience for a station in one market can be vastly larger than the audience in another. This does not, of course, guarantee that large market stations will have large audiences, especially because large markets tend to have more media outlets. Nevertheless, it sets the stage for bigger audiences and bigger audience revenues. In fact, before ratings data were available, radio stations would often sell time on the basis of their market size. Even today, cable networks will occasionally represent their audience in terms of potential viewers.

Potential audiences, however, are not just a matter of the number of people living within reach of a signal. The composition of the population can have an important impact on long-term patterns of exposure as well. In 1963, Gary Steiner published a well-known audience survey entitled, *The People Look at Television*. In 1973, and again in 1985, Robert Bower published follow-up surveys. Among the many areas of investigation were underlying changes in the U.S. population itself. Table 7.1 summarizes some of these changes.

There have been shifts in the relative size of white collar and blue-collar populations, in the number of people under 30 years old, and most notably in the level of education throughout the population. Occupation, age, and education may all be associated with an appetite for certain types of programming, and to the extent that the size of different segments of

TABLE 7.1
Trends in U.S. Population Demographics[a]

	1960	1970	1980
Age[b]			
18 to 29	23.2%	28.0%	31.0%
30 to 49	40.7	35.0	33.2
50 and older	36.0	37.1	36.0
Education[c]			
Less than high school	56.5%	40.1%	33.7%
High school graduate	26.7	35.7	34.4
Some college	16.8	24.2	31.9
Occupation[d]			
White collar	43.1%	48.6%	53.0%
Blue collar	56.9	51.4	46.9

[a] Adapted from Bower (1985).
[b] Based on persons 18 and older.
[c] Based on persons 25 and older.
[d] In 1960 and 1970, based on persons 14 and older, in 1980 based on persons 16 and older.

the population increase or decrease, the audience for those program types might be expected to vary as well. A far-sighted media operator will take population shifts, most of which are quite predictable, into account when planning for the future.

The recent rise in Spanish language programming can also be viewed, at least in part, as a result of newly emerging potential audiences. In 1970, Latinos or Hispanics accounted for 4.5% of the U.S. population. By 1980, that segment of the population had grown to 6.5%. Because of immigration and relatively high birth rates, the Latino population is expected to double by the turn of the century. The same rapid growth rates characterize the U.S. Asian population. Such demographic changes provide new markets for advertisers and the media, and may help explain corresponding changes in media use.

The second structural attribute of audiences, and one of the most powerful determinants of exposure to the electronic media, is *audience availability*. Although potential audiences set an absolute physical limit on audience size, our daily routines set a practical limit on how many people are likely to be using either radio or television at any point in time. It is widely believed that the number of people using a medium has little, if anything, to do with programming, and almost everything to do with who is available. Most practitioners take the size of the available audience as given, just as they would the size of the population itself. In practice, the available audience is most often defined as the number of people using a medium at any point in time.

Audience availability is a slipperier concept than you might imagine, and it is worth a brief digression. Measuring audience availability by audience size is partly a matter of convenience and partly a matter of definition. Although there is certainly not a one-to-one correspondence between availability and use, the difference is critical only to the extent that non-use is attributable to programming. For example, a person who is at home, but never even thinks about watching television is, for all intents and purposes, unavailable. Therefore, the only genuinely available audience that is missing from the total audience is one composed of people who are aware of the available programming and not viewing for that reason. Our suspicion is that these constitute only a small fraction of the available audience. In any event, we treat audience size as a measure of availability. We also have more to say about potential and available audiences in chapter 8.

The idea that the total audience is determined by things other than the programming is common to both the conventional wisdom of programmers and to more formal theories of audience behavior. In 1971, Paul Klein, then a researcher at NBC, offered a tongue-in-cheek description of the television audience. Struck by the amazing predictability of audience size,

Klein suggested that people turned the set on out of habit, without much advance thought about what they would watch. After the set was on, they simply chose the least objectionable program (LOP) from available offerings.

In effect, what Klein suggested was that audience behavior is a two-stage process in which a decision to use the media precedes the selection of specific content. The tendency of people to turn on a set without regard to programming is often taken as evidence of a *passive audience*. Whether such passivity is good or bad is debatable. But laying value judgments aside, this process does seem to characterize much of our media use. The conceptual alternative, a thoroughly *active audience,* appears to be unrealistic. Such an audience would turn on a set whenever favorite programs were aired, and turn off a set when they were not. We know, however, that daily routines (e.g., work, sleep, etc.) effectively constrain when we turn sets on. We also know that many people will watch or listen to programming they are not thrilled with, rather than turning off their sets.

Of course, this is a broad generalization about audience behavior. It is not intended to rule out the possibility that people can be persuaded to turn their sets on by media content. Major events like the Super Bowl, or dramatic news stories undoubtedly attract people to the media who would not otherwise be there. Nor does it mean that everyone is the same in their approach to media. Activity levels probably vary from person to person. Even the same person might be choosy at one time and a "couch potato" the next. Overall, however, this two stage process appears to explain audience behavior rather well.

Whatever the explanation, the size of the audience, like other forms of mass behavior, is quite predictable. Three patterns are apparent: seasonal, daily, and hourly. Seasonal patterns of media use are most evident in television viewing behavior. Nationwide, television use is heaviest in the winter months of January and February. Nielsen reports that the average household has a set in use for almost 8 hours a day at this time of year. During the summer months, household usage drops to about 6.5 hours. This shift seems to occur because viewers have more daylight in the summer and pursue outdoor activities that take them away from the set. Such seasonal changes have the overall effect of depressing HUT levels in the summer, and boosting them in the winter. Underneath these household-level data, however, different segments of the population may exhibit unique patterns. For example, when school is out, daytime viewing among children and teenagers soars. Seasonal differences might, similarly, vary by region of the country.

Audience size also varies by day of the week. Nationwide, prime-time television audiences are higher on weekdays and Sunday, and lower on

Fridays and Saturdays. The late-night audience (e.g., midnight) on Friday and Saturday, however, is larger than it is during the rest of the week. This, too, seems to reflect a change in people's social activities on the weekends. Radio audiences also look different on weekdays and weekends. On Saturday and Sunday, the early morning drive-time audience is reduced. We might also point out that early in the week, radio audiences are often lower than toward the end of the week. As we noted in the chapter 5, however, this can probably be attributed to diary fatigue, rather than a real reduction in early week radio listening.

The most dramatic shifts in audience availability, however, occur on an hourly basis. It is here that the patterns of each day's life are most evident. Figure 7.2 is based on RADAR data, and it depicts the size of the radio audience at various times during the day, Monday through Friday. It also indicates where listening occurs: at home, in an auto, or elsewhere. As you can see, the size of the total audience increases very rapidly from about 5 a.m. to 7:30 a.m. During the morning hours, much of that listening occurs in cars as people commute to work, hence the name—drive time. It is during this time period that a radio station can typically capture its largest audiences, so it may devote considerable resources to programming. Other stations, of course, are doing the same thing, and competition for the drive-time audience can be intense. Throughout the rest of day the audience gradually shrinks in size, with much listening in the workplace. At about 2:30 p.m. the audience picks up again, making what is called the afternoon drive-time audience. Thereafter, it trails off as people return home and television begins to command their attention.

Figure 7.3 represents the size of the television audience on an hourly basis. In some ways, it is the mirror image of the radio audience. Here, the early morning audience is relatively small. Throughout the day, however, it begins to grow. At about 5 p.m., when people arrive home from work, sets go on and HUT levels rise sharply. The total size of the audience peaks between 9 p.m. and 10 p.m., which marks the height of prime-time—a peak that you can see is slightly depressed in summer. As we noted in the first chapter, it is during prime-time that the broadcast networks are able to charge a premium for commercial time. It is also during this time period that networks air the most expensive programming. The competition is stiff, but the rewards for winning a healthy share of this large audience can be substantial.

Audience availability is an important key to evaluating gross measures of audience size. A rating during one part of the day might be quite acceptable, and during another, a disaster. For example, the networks have long vied for the early morning television audience with news programs like the "Today Show" and "Good Morning America." The "winner" of these competitions will have an average rating of about 4 or 5. During

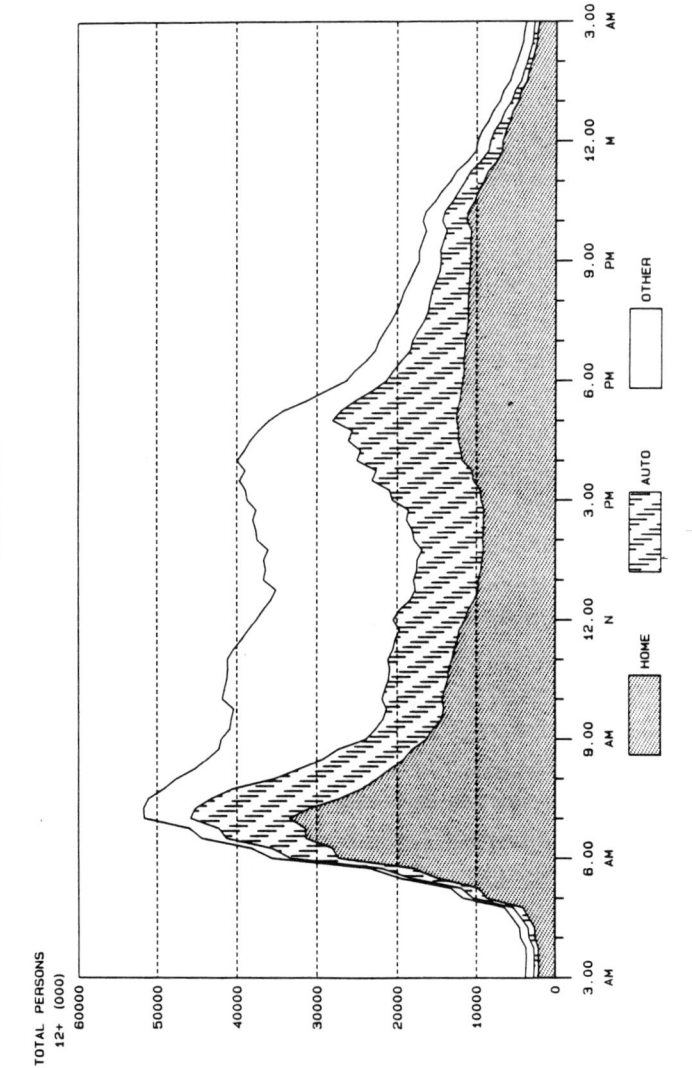

FIG. 7.2. Hourly variation in radio audience size (RADAR® 39, Spring 1989 Copyright © Statistical Research, Inc.).

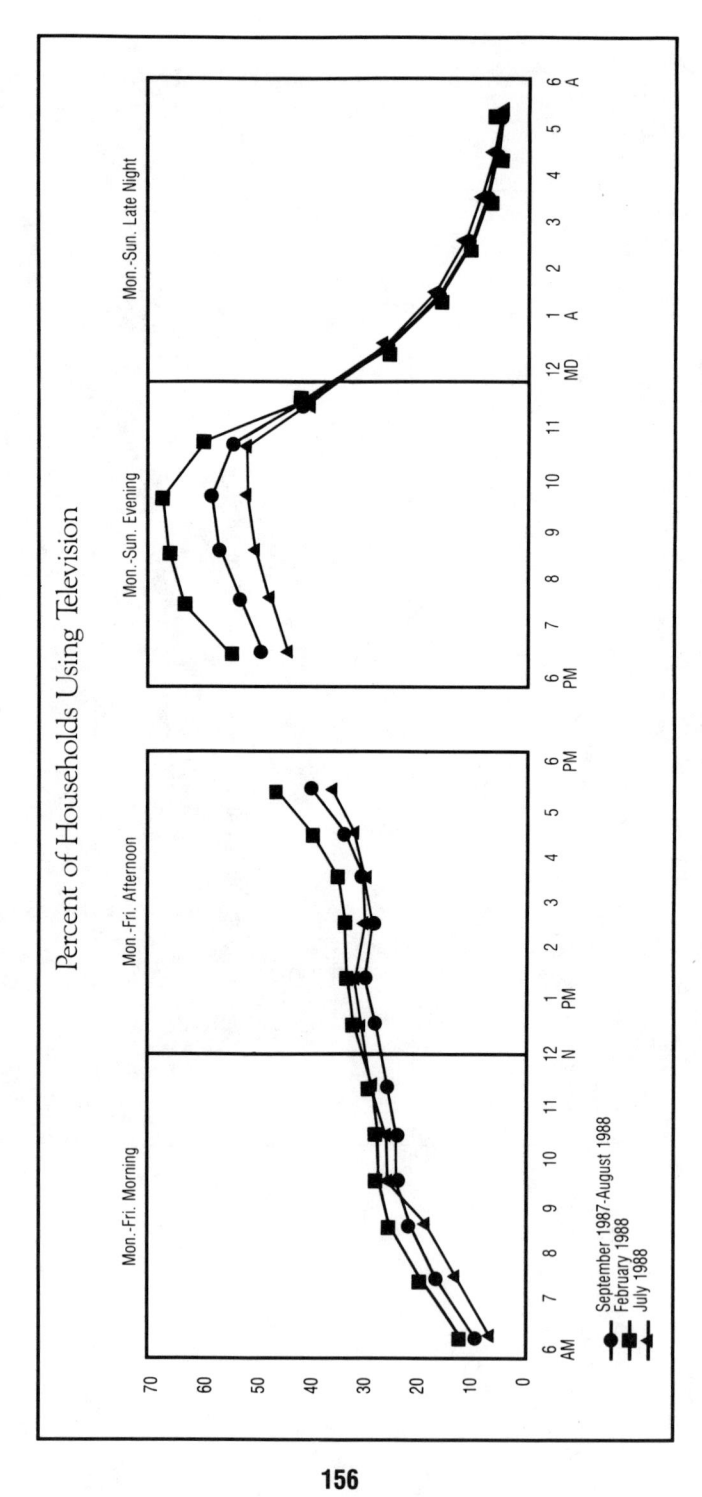

FIG. 7.3. Hourly variation in television audience size (reprinted with permission of Nielsen Media Research).

A FRAMEWORK FOR ANALYSIS

prime time, however, network shows with a rating twice that large will almost certainly be cancelled. In other words, the size of the available audience imposes certain expectations about how large an audience any one competitor can capture at any one point in time.

This fact of life is reflected in the other major measure of gross audience—audience shares. Shares, of course, express the size of a program or station's audience in a way that adjusts for changes in the total size of a medium's audience. It tells you how much of the pie you have, without reference to the size of the pie itself. A major television network, for example, tries for a share of audience in the upper 20s or better. In other words, if it can consistently attract 25% to 30% of those who are viewing, it is doing quite well. During the early morning news that means a rating of 5 or 6. During the height of prime time, however, it means a rating in the high teens. As the electronic media become more competitive, as cable services and independents lay claim to a piece of the audience, these network share expectations are likely to trend downward.

Thus far, we have considered availability only in terms of gross measurements. We know how many people are likely to be using a medium at any point in time, but we should also consider patterns of availability across time. For example, are different segments of the population differentially available? Do the people who watch television at 8 p.m. on one night watch again at the same time the following evening? Here too, there are some fairly stable, if less well-known, patterns to report.

The first point to remember is that different people consume very different amounts of radio and television. The average American watches about 4 hours of television a day, and listens to the radio for roughly 3 hours. These, of course, are only averages. Some people use a medium much more than the average, others much less. Indeed, about one-fifth of all Americans watch nearly 12 hours a day whereas another one-fifth average less than a half hour. A portion of this variation can be associated with the demographic characteristics of the audience. For example, people 45 or older use radio less and television more than the rest of the population. Slightly heavier than average television viewing is also associated with lower levels of income and education. These individual differences in media consumption are often explained in terms of availability. For instance, older people are thought to watch more TV because they have more time on their hands and fewer activities that take them outside the home. In any event, individual differences in media use produce certain patterns of audience overlap that are not apparent in the gross measures of audience size we have just reviewed. To understand this, imagine that the 6 p.m. HUT level is 50% on two different nights of the week. To what extent do the people who watch TV one night watch again the following night? One possibility is that the same 50% of the population is watching on both days. That would happen if half the population always watched TV and the other half never watched. Such mass behavior seems rather unlikely. Another possibility is that every one watches the same amount, and that their presence in the audience varies randomly from night to night. If that were the case, only 25% of the entire audience would have been watching TV on both evenings (i.e., 50% X 50% = 25%). In other words, audience overlap would be exactly proportional to the total audience (i.e., half the 50% HUT is made up of viewers from another night). We explore the arithmetic of this kind of chance occurrence in the last chapter. But for the moment, just recognize that this outcome is also unlikely, because we know that some people are relatively heavy users, and are therefore more apt to be in the audience on a day-to-day basis.

When comparisons of total audience overlap are actually made, the level of overlap does, in fact, exceed chance. According to Barwise and Ehrenberg (1988), observed levels of overlap tend to be one fifth higher than chance. In our example, then, roughly 30% of the population could be expected to have been using television at 6 p.m. on both evenings. Stated differently, 60% of the people who watched one night are watching again the following night. As it turns out, these patterns of overlap are stable even over more widely separated days.

One important exception to the Barwise-Ehrenberg one-fifth rule-of-thumb applies when we look at audience overlap in adjacent time periods. If we consider audience consistency hour to hour, we find the same people are much more likely to be in the audience. For example, of the people who watch television at 7 p.m. about 75% to 80% will be watching at 8 p.m. on the same night. As hours become more widely separated, however, audience overlap drops to the more usual levels.

Just as knowing patterns of audience size tempered our interpretation of ratings, an appreciation of audience overlap can inform our assessment of cumulative measurements. For example, among segments of the population that are heavy users of television, we should expect a relatively high frequency of exposure to advertising messages. Similarly, if we are examining audience flow between programs that are scheduled back to back, we should expect higher levels of audience duplication than between programs that are scheduled days, or even hours apart. We discuss these patterns of audience flow in more detail later.

Thus far, our approach to explaining exposure has had almost nothing to say about people's preferences, or the appeals of different kinds of programming. Remember, however, that we have characterized audience behavior as a two-stage process. Turning on a set may have little to do with specific content, but once a decision to use the media has been made, people's likes and dislikes, as well as a number of other factors, do play a role. These factors are the microlevel determinants of audience behavior.

Individual-Level Determinants. The most important individual level determinants of exposure to programming are, broadly speaking, people's preferences. Much of a programmer's skill in building an audience comes from an ability to judge what people will or will not like. This strategy for explaining audience behavior is also popular with academics from a variety of disciplines including marketing, economics, and social psychology. Each group of theorists has a slightly different frame of reference, and each employs a slightly different vocabulary. We briefly review these theoretical perspectives. But before we do, we offer a few general observations about this way of explaining behavior.

Preferences are conditions of mind that cannot be directly observed. Their "invisibility" is important because it means we can know of their existence only indirectly. Strict behaviorists are skeptical of such approaches, preferring instead to study only that which can be seen. Most of us, however, find ideas like preferences, attitudes, values, needs, tastes, and wants useful tools for understanding human behavior. We use these concepts in our everyday lives. But how can we really know a person's preferences? How do we even know such things exist?

Often, we infer preferences from a person's behavior. This is the approach taken by economists, as expressed in the axiom of "revealed preferences" mentioned in chapter 3. It is also common in psychological research on attitudes. Basically, there is an expectation that a person's attitudes will be consistent with his or her behavior. In fact, this assumption is inherent in the definition of *attitudes* as "predispositions to respond in a particular way toward a specified class of objects" (Rosenberg & Hovland, 1960, p. 1). This simple link between affect and action offers a powerful concept for explaining behavior that typifies the approaches cited here.

Economic theory presents a model for explaining program choice. Peter Steiner (1952) is credited with the seminal work in this field. Steiner and those who have extended his work (e.g., Owen, Beebe, & Manning, 1974) take the approach that a person's choice of programming is analogous to his or her choice of more conventional consumer products. Hence, older theories of product competition have served as the model for economic theories of program choice. Under such theories, two rather important assumptions about the audience are made. First, it is assumed that there are "program types" that can be defined in terms of audience preferences. Second, when considering advertiser-supported programming, it is assumed that programs are a "free good" to the audience member.

Program type is a concept with which we are all familiar. In television, we talk about soap operas, cop shows, and situation comedies. In radio, we describe station formats as contemporary hits, country, or all news. These are the familiar industry categories, but we can actually define program types in any number of ways. For example, programs could be grouped as entertainment or information, adult or children's, color or black and white, first-run or rerun, and so on. To stipulate that a program typology must be "defined in terms of audience preferences," as the economists do, forces us to consider exactly which categories of content are systematically related to audience likes and dislikes. In theory, such a typology must mean that people who like one program of a type will like all other programs of that type. Conversely, people who dislike a program, must dislike all others of that type.

We see anecdotal evidence of such reasoning in the operation of the broadcasting industry. Popular movies are made into television series of the same sort. Hit TV programs are imitated the following years, all apparently on the assumption that there is an audience out there who likes that kind of material. As one pundit put it, in television nothing suceeds like excess. Marketing researchers have conducted more formal studies to identify the content characteristics that seem to polarize people's likes and dislikes. There is good reason to question some of the methods these analysts have employed, but what is generally discovered is that common sense industry categories come as close to a viewer-defined typology as anything. That is to say, the people who like one soap opera do, in fact, tend to like other soap operas, and so on. Although there is relatively little analogous work on music preferences, similar patterns of preference for rock and roll, country music, or other types are quite easy to imagine.

One additional, rather interesting facet of people's program preferences has emerged from this line of research. People's dislikes are more clearly related to program type than are their likes. In other words, what we like may be rather eclectic, but what we dislike is more readily categorized. You might test yourself on this point by writing down the five TV shows you like most, and the five you like least. For some people, it is hard to express dislikes in anything other than program types.

The second assumption of economic models of program choice, that programs are a free good, requires further consideration. In the process of stating the assumption, the "opportunity cost" of audience time, and the increased costs of advertised products are usually acknowledged. What is implicit in the assumption, however, is often left unsaid. Remember that programs have been likened to consumer goods. In the absence of any price for programs, it seems logical that the only thing left to explain audience choice is preference. The assumption that preference is a cause of choice is certainly in keeping with the other economic theories, and is much the same as the psychologist's expectation of attitude-behavior consistency.

The economist's models of program choice differ in how they resolve the active-passive question raised by research on audience availability. Steiner (1952) assumed a thoroughly active audience in which audience size was determined by the presence of people's preferred program types. According to Steiner's model, when your favorite program type wasn't on, neither was your set. Subsequent models, however, have relaxed that rather stringent assumption, and adopted a two-stage process, much like that proposed by Klein.

Selective exposure theory offers another way to explain people's use of media content. It has been developed by social psychologists who, among other things, are interested in understanding the media's effect on audience members. In its earliest form, selective exposure theory assumed that people had certain attitudes, beliefs, or convictions that they were loath to change. These predispositions led people to seek out communications that were consistent with their beliefs, and avoid material that challenged them. Simply put, people were thought to "see what they wanted to see," and "hear what they wanted to hear."

This common sense notion gained greater credibility in the 1950s and 1960s with the introduction and testing of formal psychological theories like cognitive dissonance. Early studies seemed to indicate that people did select media materials in such a way as to support their existing belief systems or "cognitions." Hence, selective exposure to news and information appeared to be an important principle in understanding an individual's choice of programming.

By the 1970s, however, more exacting studies began to cast doubt on the lock-step nature of selective exposure to information. Although research in this area languished for a while, more recent, broader, variations of selective exposure theory have been introduced. For example, experimental studies by Zillmann and Bryant (1985) have shown that people's choices of entertainment vary with their moods and emotions. Excited, or over stimulated people, are more inclined to select relaxing program fare, whereas people who are bored are likely to choose stimulating content. Emotional states, in addition to more dispassionate cognitions, all seem to influence our program preferences.

Gratificationist theory provides a closely related, if somewhat more comprehensive, perspective on audience behavior. Studies of "uses and gratifications," as they are often called, are also the work of social psychologists. This approach emerged in the early 1970s, partly as a reaction against the field's apparent obsession with media effects research. Gratificationists argued that we should ask not only "what media do to people," but also "what people do with the media." The research agenda of this approach was spelled out by Katz, Blumler, and Gurevitch (1974). According to them, gratificationists:

are concerned with (1) the social and psychological origins of (2) needs, which generate (3) expectations of (4) mass media or other sources, which lead to (5) differential patterns of media exposure (or engagement in other activities), resulting in (6) need gratifications and (7) other consequences, perhaps mostly unintended ones. (p. 20)

Since the early 1970s gratificationist research and theory has attracted considerable attention. Of central importance to our discussion is the gratificationist's approach to explaining "patterns of media exposure." Under this perspective, those patterns are determined by each person's expectations of how well different media or program content will gratify their needs. Such needs might be short-lived, like those associated with mood states, or relatively constant. In any event, it seems likely that the gratifications being sought translate rather directly into preferences for the media and their content.

Gratificationist theory, therefore, has much in common with economic models of program choice and theories of selective exposure. All of them cast individual preferences, however they have emerged, as the central mechanism for explaining exposure. Grandiose theories aside, this view of audience behavior also has a great intuitive appeal. Why does the audience for a hard rock radio station tend to be young men? Because that is the kind of music they like. Why do males watch more televised sports than females? Because they like it more. Figure 7.4 summarizes various well-established associations between audience demographics and pro-

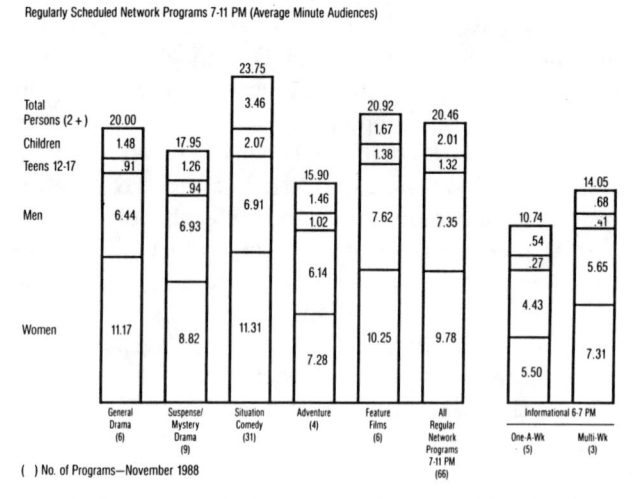

FIG. 7.4. Audience composition by selected program type (*Note:* Reprinted with permission of Nielsen Media Research).

gram type. Right or wrong, these associations are often explained as the result of differential program type preferences across segments of the population.

Unfortunately, the power of preferences to determine exposure to the media is not as absolute as many have assumed. We all know from our own experience that we often see and hear programming we do not particularly like. As in other areas of research on human behavior, we have discovered that attitudes and behaviors are not always consistent. Indeed, a disparity between people's program preferences and their program viewing is one of the most remarkable and persistent findings of the Steiner-Bower audience surveys mentioned earlier. Perhaps the most obvious explanation for these inconsistencies are the patterns of audience availability we reviewed earlier. Although many theorists tend to ignore this factor, outside the laboratory it effectively constrains program choice. Another part of the explanation can be found in the remaining microlevel audience factors.

Most research and theory on the relationship between preference and choice focuses on the individual, and assumes that personal preferences can be freely exercised in the selection of programming. This is quite typical of research in laboratory settings. Much of our media use, however, is done not in isolation, but in the company of others. This is especially true of television viewing. *Group viewing* is a rather common phenomenon, even today, when most households have more than one set. Table 7.2 summarizes the incidence of group viewing that Bower found in 1970 and 1980. The most typical group configuration is the entire family viewing together, followed by husband and wife together.

What little research there is on the dynamics of group viewing suggests that negotiation among competing preferences is quite usual. Different members of the family seem to exercise more or less influence at different times of the day. For example, programmers make much of the fact that

TABLE 7.2
Typical Viewing Groups in Single and Multiset Homes[a]

Viewing Group	Single Set 1970	Single Set 1980	Multiset 1970	Multiset 1980
Entire family	55%	41%	34%	36%
Husband and wife	17	25	26	26
Children	13	16	26	22
Mother and child	9	9	5	6
Father and child	4	8	7	5
Other	3	1	3	5
Base sample (100%)	443	185	613	597

[a] Adapted from Bower (1985).

children are often in control of the television set in the late afternoon when they return from school. Exposure to television programming, then, results not only from who is available and what they like, but who is actually making the program selections. People get their first choices some of the time, but can be "out voted," at other times. Even if they are overruled, however, they will often stay with the viewing group. Ask any parent of a young child whether they are watching more "Sesame Street" since the child's arrival. Ask the child if he or she sees more of the evening news than he or she would like. In effect, some of our exposure to media is enforced, in spite of our preferences. Group viewing, or for that matter, group radio listening in the car, can constrain the relationship between preference and choice.

The last audience factor to enter the picture is *awareness*. By awareness, we mean a knowledge of the programming that is available to you. Much theorizing about the audience presupposes perfect awareness on the part of audience members. In other words, program selection is assumed to occur with a full knowledge of programming options. Although that assumption might be workable on a very abstract level, or in very simple media environments, it does not seem to work well in the media-rich environments that confront most audience members.

If, as is sometimes the case, people select programming without a full understanding of their options, the interpretation of choice as an expression of preference is complicated. How often have you missed a show you would have enjoyed because you unthinkingly happened to watch something else? Have you ever "discovered" a favorite program or station that had been on the air for some time? As more and more programming services compete for the attention of the audience, these sorts of breakdowns between preference and choice are likely to be increasingly common.

Audience awareness is an intriguing concept—one that has received relatively little attention from the audience researchers. How, for instance, do people become aware of their options? They sometimes rely on newspaper listings or television guides. They might get a word-of-mouth recommendation from a friend. With the advent of remote control devices, they may simply "graze," or skip from channel to channel. What kind of awareness does that produce? Is a program title or a brief snippet of programming enough to make an informed judgment of what to view or listen to? If, as Paul Klein (1971) suggested, it is an avoidance of objectionable programming that produces program choice, how do people make those determinations? What characteristics brand a show as objectionable?

These questions have gone largely unanswered, and should concern academics and practitioners alike. We do know that one of the tricks to

A FRAMEWORK FOR ANALYSIS

operating a successful radio station, particularly in a competitive market, is to develop a clear image for the station. That way, fans of a particular format will know where to find that kind of programming. We also recognize that an effective television programmer must be able to not only pick appealing programs, but to promote them as well. The more choices the audience has, the more important these efforts at positioning and promotion will become, and the more we could use a systematic body of knowledge on audience awareness.

The role that audience preferences play in determining audience behavior, then, is far less tidy than we might wish. A small rating might indicate that people do not like a particular station or program. It might also indicate that they simply did not know it was there. Further, the audience factors we have reviewed are only half the picture. The structure of the media themselves have an impact on patterns of exposure.

MEDIA FACTORS

As with audience factors, media factors can be grouped as structural or individual. The structural attributes of the media complement the structural features of the audience. They include market conditions and how available content is organized. Individual level media factors vary in tandem with individual audience attributes, defining differences in the media environment from household to household.

Structural-Level Determinants. The first structural characteristic of the electronic media that should be taken into account is *coverage,* which is the extent to which people are physically able to receive a particular channel or medium of programming. In the United States, the universal availability of both radio and television is usually taken for granted. In other countries, especially in developing nations, universal coverage is not the rule. Even in the United States, newer forms of media are not available to all households. Table 7.3 summarizes the growth of various electronic media in this country.

Obviously, a medium's coverage of the population has a powerful impact on its ability to attract audiences. Early television audiences had to be small because few people had receivers. Similarly, cable television's audiences are shaped, in the first instance, by the fact that only about half of U.S. households subscribe to that medium. Indeed, barring some major change in the technology and/or regulation of cable services, its coverage will always be more limited than the broadcast networks.

Although the presence of a television or radio set in the home suggests something about how widely a medium covers a population, it reveals

TABLE 7.3
Growth of Electronic Media in the United States (1960–Present)[a]

Year	Total U.S. Households[b]	% with Radio	% with Television	% with Cable	% with Pay-Cable	% with VCR
1960	53	96	87	—	—	—
1965	57	98	93	2	—	—
1970	63	98	95	6	—	—
1975	71	98	97	12	—	—
1980	81	98	98	19	6	1
1981	81	98	98	22	10	2
1982	83	98	98	29	15	3
1983	85	98	98	33	19	5
1984	85	98	98	39	23	10
1985	87	98	98	42	25	20
1986	88	98	98	45	26	35
1987	89	99	98	47	24	48
1988	90	99	98	48	26	57
1989	92	99	98	52	28	63
1990	94	99	98	55	29	67

[a] *Sources:* Television Bureau of Advertising (1990) and Radio Advertising Bureau (1990).
[b] In millions

very little about the quality of coverage. Each medium is made up of more discrete units like stations, channels, and networks. A person with a radio receiver might be able to receive one local station and not another. A cable subscriber might have access to only a portion of the cable networks now in operation. From each station or network's point of view, this produces a pattern of coverage that circumscribes the audience it can attract. These channel specific patterns of coverage result from a combination of factors.

Among broadcast stations, the most important determinant of coverage is the signal. The neatly drawn concentric circles that show up on a map of a station's coverage area are deceptively simple. FM stations, for example, emanate radio waves that travel in a direct, line-of-sight, manner. As a result, people who live in a valley may be unable to receive a signal, whereas others living further away can. The vagaries of geography can also affect the coverage of a television station. AM radio, on the other hand, has the ability to radiate a signal that can be received over great distances. *Sky waves,* as they are called, bounce off the ionosphere at night creating a secondary coverage area. The stations licensed to "clear channels" can sometimes be received a continent away. The pattern of this secondary coverage area, however, is unpredictable.

Even if a receiver can tune to a particular signal, the quality of the transmission may somehow handicap the station. For example, AM radio stations are more susceptible to static interference than are FM stations.

They also have less fidelity, and so are less satisfactory for broadcasting music. UHF television stations operate on higher frequencies than VHF stations, and therefore, have signals that "attenuate," or diminish more rapidly. They may also be handicapped by the presence of old-fashioned sets and receiving antennas in people's homes, which make reception more difficult and tuning in less likely. Each of these factors has a practical impact on a station's coverage, and hence, its ability to attract an audience.

Networks encounter different problems of coverage. As we saw in chapter 1, the major broadcast networks reach nationwide audiences through their affiliates. With the exception of a few stations that are actually owned and operated (O & Os) by the networks, affiliates are independent businesses that act in their own self-interest. This means that an affiliate may not carry all of a network's programming if it believes some other programming strategy will be more profitable. A commitment on the part of a station to hold its schedule open for a network-fed program is called *network clearance*. Usually, of course, an affiliate will clear its network's programming. Sometimes, however, a network program, particularly one that is unpopular, will not get clearance on all affiliates. If this happens on enough affiliates, or even a few affiliates in larger markets, it can seriously erode the network's coverage, compounding the problem of a low rated show.

In addition to routine matters of network clearance, an affiliate may decide to preempt a show that it ordinarily carries. In fact, even O & Os occasionally preempt a network feed. Preemptions also reduce network coverage. All these potential deviations from straightforward network coverage complicate both the production and interpretation of network program ratings.

Other competitors for the national audience face similar problems of coverage. Like networks, barter syndicators must build coverage on a market-by-market basis. We have also noted that all cable services run into an absolute limit on coverage imposed by the medium itself. Even within that limit, individual cable networks must struggle to achieve carriage on thousands of different cable systems around the country. Although major cable networks, like CNN or ESPN come close to reaching the cable universe, others have spottier coverage. Sometimes individual cable systems simply do not have the channel capacity necessary to carry more services. Sometimes a system owner, especially one with financial interests in programming services, will decline to carry competing services.

All these aspects of program or network coverage impose real physical limits on what programming is available to which audiences. For that reason, they are powerful determinants of audience size. But the structural attributes of the media go beyond channel or program availability.

Within each channel, there is a certain sequence of programming that affects patterns of exposure. Similarly, because channels compete with one another, they effectively offer the audience a series of forced choices. Although the ability to tape and replay programming can, in principle, break this rigid structure, in practice relatively little taping is done. *Program scheduling,* within and across channels, therefore, is widely believed to be an important factor in shaping the size, composition, and flow of audiences.

The first factor to consider is the number of program options that confronts the audience. On average, the number of choices before the audience has increased. This is especially true for television. Table 7.4 depicts changes in the average number of TV channels available to American households.

Although virtually all programming services are competing for the audience, different competitive strategies are employed. If it is assumed that there is a relatively large audience for some particular type of programming, then two or more channels or stations will split that audience by offering programming of that type. This will continue to happen until that program type audience has been divided into small enough pieces that it makes sense for the next competitor to offer a different kind of programming. This is the strategy we called *counter programming*. In practice, you often see this kind of counter programming done by independent stations in a market.

When there are only a few competitors, there tends to be an "excessive sameness" about program offerings. Each tries to maximize its audience with "lowest common denominator" programming. According to the economic models of program choice we just reviewed, as the number of com-

TABLE 7.4
Growth in Average Number of Channels Receivable Per U.S. Household[a]

Year	Over-the-Air Stations	Total Channels
1950	3.8	NA
1976	7.7	NA
1981	9.1	NA
1982	9.8	NA
1983	10.3	14.6
1984	10.3	17.2
1985	11.0	18.8
1986	11.2	19.4
1987	11.4	22.4
1988	11.6	25.1
1989	12.1	30.5

[a] Based on NTI data, courtesy of Nielsen Media Research.

petitors increases, program services become more differentiated. This process continues until potential audiences are so small as to make it unlikely that one will recover the costs of providing a program service.

Presenting the audience an array of competing services has two important consequences. First, from the individual's point of view, programs are offered up as a series of forced choices. It is quite possible to encounter situations in which two desirable programs are on opposite one another. Indeed, presenting the viewer with tough choices is what the competition is all about. Had those shows been scheduled at different times, the viewer could have watched both. But given the nature of the program schedule, a choice is forced. This kind of competitive scheduling is another reason why a person's preferences may not be the best guide to actual patterns of exposure. Second, when individual choices are aggregated, an increased number of competitors will fragment the audience.

Audience fragmentation is a matter of breaking the mass audience into smaller and smaller segments. It seems to be an inevitable outcome of allowing greater competition for the audience, at least if competitors are more-or-less equally matched. In most large radio markets, for example, no one station can expect to consistently dominate the others. Each targets a segment of the audience and must usually settle for a share of audience in the single digits. The same is true in television, although as our discussion on coverage should indicate, some of the competitors have a considerable advantage. Specifically, broadcast networks are capable of reaching virtually all households, whereas cable networks can reach only half. Nevertheless, as Table 7.5 suggests, the growth of independent stations, as well as cable television, has gradually eroded the network's total share of prime-time audiences.

TABLE 7.5
Trends in TV Network Audience Shares[a]

Year	Total Network Daytime Share	Total Network Prime-Time Share
1976–1977	81	91
1980–1981	73	84
1981–1982	68	80
1982–1983	65	77
1983–1984	65	75
1984–1985	62	73
1985–1986	61	73
1986–1987	58	71
1987–1988	59	66
1988–1989	57	64

[a] Based on NTI data. Provided by the Cabletelevision Advertising Bureau.

Just as a knowledge of macrolevel audience factors shaped our expectations about audience ratings, these structural attributes of the electronic media should temper our interpretation of gross audience measurements. More competitive marketplaces almost inevitably mean smaller ratings and shares for each of the competitors. Not all competitors, however, are created equal. Consider the technical attributes of each. Some handicaps can be overcome by clever engineering, but others, like expanding coverage areas may be more intractable. You can get a sense of how different levels of competition affect audience size by comparing the viewing of households that do and do not have cable television. Table 7.6 presents such a comparison.

The left-hand column summarizes the viewing of homes that do not have cable television. By definition, they are restricted to watching only those signals that they can receive over the air, including affiliates, independents, and public TV stations. Under such circumstances, network affiliates command 73% of all the television viewing done in the home. In the second column is a summary of viewing in basic cable homes. Here, programming originated by cable systems accounts for 24% of all viewing. The affiliates' share of audience drops to 56%. The third column represents the viewing of homes in which at least one pay cable service, like HBO,

Table 7.6
Weekly Share of Television Viewing Across Broadcast, Cable, and Pay Cable Households[a]

Programming Source	Broadcast (42%)	Cable (27%)	Pay Cable (31%)	All TVHH (100%)
Affiliates	73	56	45	59
ABC	25	18	16	20
CBS	25	19	15	19
NBC	24	19	15	19
Independents	27	21	21	23
Superstations	5	10	9	8
Local	21	8	9	13
Distant	2	3	2	2
Cable	—	24	42	21
Basic	—	24	24	15
Pay	—	—	18	6
Public	5	4	2	3
Average Weekly Hours of Viewing (100%)[b]	44:33	47:32	57:01	49:13

[a] Based in NTI data (April 1989). Provided by the Television Bureau of Advertising.
[b] Sum of parts may exceed various totals due to simultaneous viewing in households.

is available. Combined with basic cable use, these services account for over 42% of all viewing, and leave the affiliates with a 45 share. Certainly, no one cable service is likely to match the audience of a network, but in combination, they do take a toll.

The structural attributes of media are also associated with well-established features of audience duplication. *Audience duplication* is the extent to which two different programs have an audience in common. When those programs are scheduled in adjacent time periods, duplication is usually referred to as *audience flow*. Patterns of audience flow can certainly affect the size of any one program's audience, but the study of audience duplication involves cumulative measurements. We describe the techniques used to assess audience duplication in chapter 9, but for now we concentrate on what we know about these patterns.

The most widely known pattern of audience flow is called an *inheritance effect,* which occurs between programs that are scheduled back-to-back on the same channel. When one program ends, an unusually large percentage of its audience will stay tuned to the following program. That is, there is an especially high level of audience duplication between adjacent shows on the same channel. The available research suggests that, on average, about 50% of those who view the first program in a line-up will watch the program that follows it. Beyond adjacent programs, however, inheritance effects diminish rather rapidly.

Obviously, if the first, or "lead-in" program has a large audience, the following program stands to benefit. This feature of audience flow has given rise to a number of television programming strategies that we described in chapter 2 (e.g., hammocking, tent-poling, etc).

Inheritance effects seem to occur, in the first instance, because of patterns in audience availability. We have noted that audience overlap is high in adjacent time periods. In should not be surprising, therefore, to discover high audience duplication between back-to-back programs. In other words, an 8 p.m. show tends to have the same audience as an 8:30 p.m. show, because the same people tend to be in the television audience at both times. Beyond audience overlap, however, other factors effect the level of audience inheritance. The most important of these is the number of competing programs that are starting in the break between the adjacent pair. If the other major networks are in the middle of programs when the first show ends, audience flow to the second program is quite pronounced. If viewers have a number of choices during the break, the level of inheritance diminishes.

Theoretically, the kind of programs that are scheduled back-to-back should also affect the level of inheritance. That is, because people are presumed to have consistent preferences for programs of a type, following one show with another of the same type should be more effective in retaining the lead-in audience than scheduling very different types. Although that appears to be true, program type does not have a powerful impact on levels of inheritance. Part of the problem may have to do with knowing which programs are truly of a type. Another difficulty is that in the real world, vast differences between adjacent programs usually do not occur. No programmer is likely to schedule "Mighty Mouse" after the "Metropolitan Opera". That means that the effect of program content is difficult to isolate. It is worth remembering, however, that the one factor clearly tied to program preferences is a rather weak determinant of audience inheritance.

A second feature of audience duplication is *channel loyalty*. Like inheritance effects, this is a predictable kind of audience duplication between different programs on the same channel. In this case however, the programs can be scheduled days apart. The level of audience duplication is not as great as that found with adjacent programs, but is still greater than chance. Many researchers have found evidence of channel loyalty, but the most extensive program of research has been conducted by Goodhardt, Ehrenberg, and Collins (1987). According to their reports, people who watch a channel one day are more likely than the general public to watch it again on another day. The level of audience loyalty varies somewhat from channel to channel. But for the major U.S. networks, it appears that the audience for one network show is 50% to 60% more likely than the population in general to watch that network on another day. This method of analysis has been codified in what Goodhardt et al. (1987) called the "duplication of viewing law," which we explore more fully in chapter 9.

Just why channel loyalty exists is something of a mystery. One explanation is that underlying patterns of audience availability produce the result. For example, channel loyalty is higher in the late afternoon, and late at night. During these time periods large blocks of the audience are consistently unavailable, thus the remaining audience is, by default, more likely to show up in program audiences on a day-to-day basis. Similarly, when we look for channel loyalty among viewers who all watch the same amount of television (i.e., have similar amounts of available time), the phenomenon seems to disappear.

One factor that does not appear to contribute to audience duplication within a channel is program type. That is, the audience for one CBS situation comedy is no more likely to watch another CBS sitcom than any other kind of program on CBS, unless of course, they are scheduled back-to-back. When we look at the audience for broadcast network television then, there is very little evidence of what might be called *program type loyalty*. This, of course, is contrary to theoretical expectations built on a simple linkage between preference and choice. The tendency to systematically like or dislike programs of a type, does not translate directly into a systematic tendency to view or not view.

Program type loyalty, however, might be evident in media environments where content is more differentiated than it is on network television. We have noted that increasing competition can produce greater content differentiation among channels. This certainly seems to be true in radio where stations typically specialize in a type of content, or format. Compared to television, listeners demonstrate marked station loyalty. Even some television audiences, like those of religious or minority language stations, exhibit high levels of duplication or loyalty. The question is, to what extent should these heightened levels of audience duplication be attributed to program type as opposed to channel?

Whichever is the case, comparatively high levels of within-channel audience duplication may become more usual with the growth of media services like cable networks. Most of these offer specialized content, such as sports and music video. These program services are likely to appeal to some segments of the audience and not to others. In fact, many are designed to do so. At the same time, as cable services, they are systematically unavailable to large blocks of the audience. Such systematic nonavailability tends to promote channel loyalty, as is the case with late-night viewers. For example, people who watch MTV once, are more likely than the general population to watch again, because half the population does not even subscribe. The possibility that these conditions could "polarize" the television audience will bear further investigation.

Another commonly studied pattern of audience duplication is called *repeat viewing*. Repeat viewing considers the level of audience duplication between different episodes in a series. For programs that are "stripped" five times a week, the episodes are only a day apart. For other series, like those in prime time, they are a week apart. Most research on repeat viewing has used diary data, and has therefore concentrated on stripped programming. These studies indicate that the typical level of repeat viewing is on the order of 50%. That is, of the people who watch a program on Monday, 50% of them watch again on Tuesday, and so on. High rated programs have a slightly higher level of repeat viewing than low rated shows. With the exception of soap operas, which have repeat viewing of about 60%, program type does not seem to effect the level of audience duplication.

More recent research using peoplemeter data has examined repeat viewing on a week-to-week basis. The results are a bit sketchier, but they indicate that repeat viewing is much lower, on the order of 25% to 30%. This may truly indicate that week-to-week repeat viewing is lower than day-to-day viewing. It may also result from methodological differences.

Because diaries are often filled out after the fact, they may reflect a recollection of consistent viewing, even if actual use was more sporadic. Additional work needs to be done before the "truth" is known.

The most plausible explanation for these levels of repeat viewing is, as you might guess, audience availability. The question is, what happened to the audience who viewed the program yesterday, but not today. Are they watching something else? Analyses of diary data indicate that almost all of them are not watching television. Hence, patterns of audience overlap seem to explain repeat viewing rather well.

We have seen how structural aspects of the media themselves interact with audience factors to produce major patterns of media exposure. Barring an unusual effort on the part of an audience member, scheduling a number of programs opposite one another precludes exposure to all but one, even if they are equally appealing. Scheduling a new program after an established hit increases the likelihood that it will be seen. Scheduling it after a poorly rated show greatly reduces the likelihood of exposure. Programs scheduled on the same channel are more likely to have a duplicated audience than programs scheduled on different channels. Scheduling programs in adjacent time slots will increase audience duplication even further. There are, however, a few micro level media factors that we should review to complete the picture.

Individual-Level Determinants. Network coverage and program scheduling are generally beyond the control of an audience member. But certain aspects of the media environment are within the individual's control. In fact, this is truer today than it has ever been. As new technologies and programming alternatives enter the marketplace, each of us has greater latitude in shaping a media environment to suit our purposes. These decisions can certainly affect our exposure to the media and are closely related to the microlevel audience factors we reviewed earlier.

Cable subscription is one such alternative for shaping a media environment. We have touched on cable often in the preceding discussions. Although much about cable's organization and availability is appropriately conceptualized as a structural variable beyond a person's control, the decision to subscribe is ultimately made by each individual household. Cable, in other words, is not just something that is done to us, it is also something we elect to do. This self-selection into the cable universe is one reason why comparisons of cable and noncable households must be made with care.

Just why people subscribe to cable varies from home to home. We do know that cable subscribers have higher incomes that nonsubscribers. We also know that cable households tend to have more people living in them.

A FRAMEWORK FOR ANALYSIS 175

This is especially true of families that buy a pay cable service. With more children, and more money to spend, subscription to cable probably makes sense. Gratificationists have pointed out that cable subscribers express a need for greater variety and control over their viewing environment. Others undoubtedly subscribe just to improve the quality of over-the-air reception.

Researchers have also observed that cable subscribers have a somewhat different style of viewing television. Confronted with a large number of channels from which to choose, cable subscribers apparently develop a *channel repertoire*. This repertoire is a subset of total number of channels available to the subscriber. The more channels there are, the larger is the repertoire. But there is not a one-to-one correspondence. Figure 7.5 depicts the relationship. As the number of available channel's increases, the proportion that are used decreases. The net result is that each cable viewer constructs an array of channels from which to choose on a day-to-day basis. This may effectively cancel out viewing on some channels, even if they can be received on the set.

Video cassette recorders (VCRs) are another microlevel factor with great potential to alter the media environment of audience members. At the beginning of the 1980s, they were virtually nonexistent as a household appliance. Today, they have surpassed cable penetration and are in about

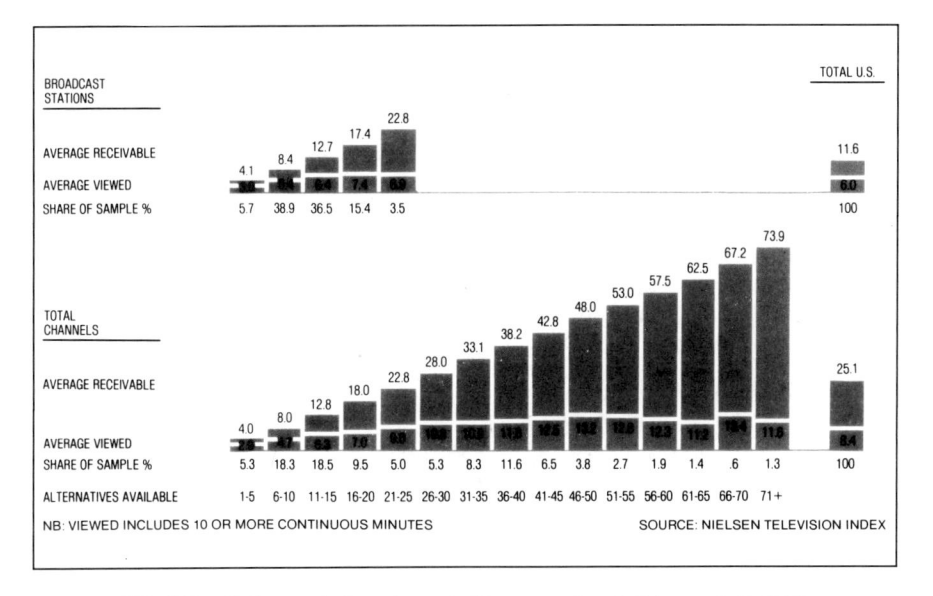

FIG. 7.5. Stations and channels receivable versus viewed, February 1–7, 1988 (reprinted with permission of Nielsen Television Index).

two thirds of all households. Many analysts have likened their adoption curve to that of color television. If that is, indeed, a model of VCR growth, their penetration into American households might ultimately exceed 90%.

VCR usage falls into two broad categories: *time-shifting,* and *library use.* As the label suggests, time-shifting involves taping a program for replay at a more convenient time. As it turns out, the lag time between taping and replay varies with how often a program is broadcast (e.g., stripped shows are replayed faster than weekly offerings). It is also quite predictable. Researchers have likened the rate of replay to a radioactive decay curve! The most taped programs are those broadcast by the major networks. Many programs that are recorded for time-shifting are never played back.

The library uses of the VCR can involve off-the-air taping. But if that is done, it is with the intention of adding the tape to a "library." Increasingly, people will buy or rent tapes for viewing at home. In fact, virtually all VCR owners report that they use their machines to show rented cassettes. The most popular rentals are major motion pictures that were successful in theatrical release. As this market grows, however, more programming made specifically for home viewing is likely to be produced.

Despite these important changes, the total amount of time that most people actually spent watching taped programming is tiny compared to the total amount of television that is consumed. For example, the time shift audience for a prime time network program rarely accounts for more that one rating point. That, of course, could change, and the impact of VCRs on patterns of exposure to electronic media should be carefully monitored.

The technology and deployment of *receivers* is yet another aspect of the media environment over which people can exercise some control. Color television sets are now in 97% of all U.S. households. Nearly two thirds of homes now have more than one television set, and over 20% have three or more sets. The location and capability of these receivers affect the quality of the media environment within the home.

Table 7.7 identifies the location of sets within the home. As one might expect, the main set tends to be in the living room or family room. But in homes with multiple sets, a large majority have a set in one or more bedrooms, some even have sets in the kitchen. As sets become smaller, and more easily portable, many will find themselves awash in television. In fact, research on the social uses of television indicates that a common function of television is to provide a kind of structured background noise.

Another innovation in television set technology with important implications for patterns of exposure to television is the remote control device. Three quarters of all television households have a set with this feature, and that number is growing rapidly, especially among main sets where

A FRAMEWORK FOR ANALYSIS

TABLE 7.7
Location of television sets in the home[a]

Location	Single Set Homes	Multiset Homes Main Set	Multiset Homes Additional Sets
Livingroom	70%	63%	5%
Family room	19	26	7
Bedroom(s)	8	6	69
Recreation or playroom	1	3	7
Kitchen	1	2	9
Dining room	1	1	1
Other room	—	—	2
Portable	—	—	1

[a] Based on 1980 data. Adapted from Bower (1985).

most viewing occurs. Because they make channel changing so easy, remote control devices strike fear in the hearts of advertisers and programmers alike. From the advertiser's perspective, viewers may be more likely to change channels when an advertisement comes on. This practice, called *zapping,* could obviously reduce exposure to commercial messages. From the programmer's perspective, audiences lost during commercial breaks, or a lull in the program itself, may be difficult to regain. The inclination of at least some viewers to change channels at the drop of a hat has been dubbed *grazing*. Whether this phenomenon becomes a major factor shaping patterns of exposure remains to be seen.

AN INTEGRATED MODEL OF AUDIENCE BEHAVIOR

Many things affect the behavior of the audience. We have defined and discussed most of these factors in the preceding sections of the chapter, but have not yet put the pieces of the puzzle together. It may be useful at this point to step back, reflect briefly on what has been presented, and then try to forge an overall framework for examining media exposure. With a comprehensive model of audience behavior, the job of summarizing, evaluating, and anticipating the data contained in the ratings will be more manageable.

Audience researchers of all stripes have devoted much time and effort to understanding people's use of the electronic media. Ad agencies and programmers have engaged in very pragmatic studies of audience formation, economists have developed rather abstract theories of program choice, and social psychologists have performed a seemingly endless succession of experiments and surveys to reveal the origins of audience behavior. Despite real progress in these, and many other, fields, there has been an unfortunate tendency for each group to work in isolation from the others. Instances of "cross-pollination" between theorists and practitioners, or even across different academic disciplines are all too rare.

At the risk of greatly oversimplifying matters, two fairly distinct approaches to understanding the audience can be identified. The first emphasizes the importance of the individual factors we discussed earlier. This perspective is typical of work in psychology, communication studies, and to some extent, marketing and economics. It also has enormous intuitive appeal, and is likely to characterize most "man-in-the-street" explanations of the audience. After all, audiences are simply collections of individuals. Surely, if we can understand behavior at the individual level, then our ability to explain larger patterns of mass behavior will follow. When we conceptualize audience behavior at the individual level, we tend to look for explanations by thinking of those things that distinguish us as individuals. Above all, we have invoked preferences (or needs and expectations) as a way to explain behavior. With this focus, however, we often miss seeing things that crystalize different levels of analysis. For instance, it was observed some years ago that "availability in any sense rarely finds a place in uses and gratifications research" (Elliott, 1974, p. 259). That is still largely true today.

The second perspective emphasizes structural factors as key determinants of mass behavior. This approach is typical in sociology, human ecology, and when applied to audiences, marketing and advertising. It down plays individual needs and wants, and concentrates on things like total audience size, coverage areas, and program schedules in attempting to understand audience behavior. Although work in this area can be highly successful in creating statistical explanations of aggregate data, it often has a hollow ring to it. One is often tempted to ask "What does this mean in human terms—what does it tell us about ourselves?" Such explanations are usually possible, but not always apparent.

It is important to recognize that neither approach is right or wrong. It is also important to note that neither approach is really complete without the other. Despite the fact that these models of audience behavior are sometimes advanced as mutually exclusive alternatives, we believe that there is much to be gained by trying to integrate them. Specifically, analyses of individual behavior might be enhanced by a more deliberate consideration of the structural factors suggested here. We know through observation that these variables are highly correlated with audience behavior, and weaving them into microlevel studies might increase the latter's power and generalizability. Conversely, research in mass behavior might be more explicit about its relationship to theoretical concepts cen-

A FRAMEWORK FOR ANALYSIS 179

tral in the individual approach. This could improve its popular acceptance and utility. It is in this spirit that we propose the following model.

A Model. We have used the term model rather loosely throughout the book. In fact, it has several meanings. In its broadest sense, it can mean an entire way of looking at the world, as implied by the term *scientific model*. It is often used interchangeably with *theory,* suggesting a tentative explanation or representation of how some phenomenon operates. It can also imply an exacting level of precision. For instance, a *mathematical* or *computer model* requires that relationships among the elements of the model be quantified. Finally, it can mean a guide or *heuristic* device, useful in organizing and thinking through phenomena of interest.

The model presented in Fig. 7.6 is a model in that last sense of the word. It is intended to help organize our thinking about audience behavior as it is defined in ratings research. A few, very broad relationships are suggested by the model, but it does not, in and of itself, provide hypotheses to be tested. It certainly falls short of being a mathematical model, although we use Fig. 7.6 as a spring board for discussing several such models in chapter 9. We should also point out that the model focuses primarily on short-term features of audience behavior.

The central component of the model, the thing we are trying to explain, is exposure to electronic media. As we argued in the beginning of the chapter, ratings analysts are interested in mass behavior, and the most common measures of that behavior can be categorized as gross or cumulative. Two large factor categories are represented as the causes of exposure:

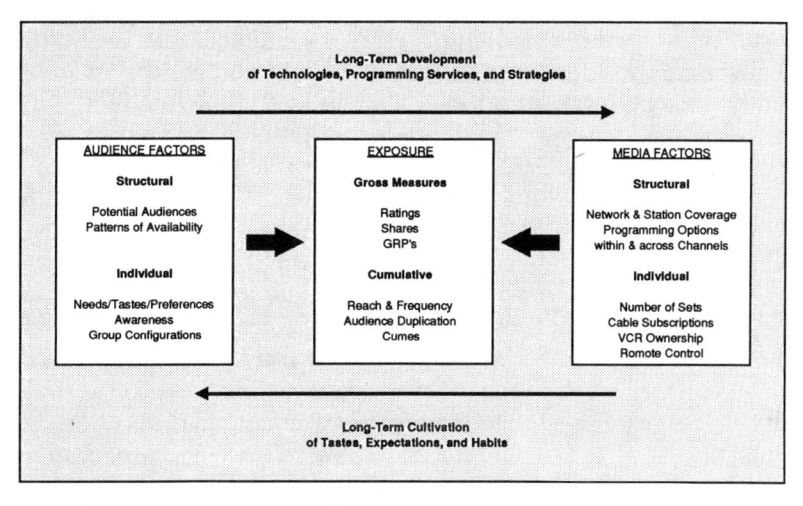

FIG. 7.6. A model of audience behavior.

audience factors and media factors. The direction of influence is indicated by arrows. For example, the model stipulates that audience factors cause ratings, not vice versa. There are also cause and effect relationships among the factors within each box. For example, audience needs probably contribute to patterns of availability, and cable subscription helps shape cable network coverage. We have opted to omit arrows suggesting these interrelationships, to keep the model a bit cleaner.

To use the model, you should identify the sort of audience behavior you wish to analyze. Are you concerned with the size of an audience at a single point in time (i.e., a gross measure), or are you interested with how audience members are behaving across time (i.e., a cumulative measure)? To begin the process of evaluating, explaining, or predicting that behavior, consider the structural determinants first. There are three reasons why we recommend you look initially for structural explanations. First, like the measures of exposure you are analyzing, they are pitched at the mass level of analysis. Second, they are more knowable. Information on program schedules, network coverage, and total audiences are typically in the ratings reports themselves. Individual factors, like audience awareness and the use of remote control devices, are a bit harder to pin down. Third, we know from experience that structural explanations work well with ratings data. If they fail to provide a satisfying answer, however, begin the process of thinking through the individual-level factors on either side of the model.

Let us work our way through a couple of examples to get a better sense of the model. Consider, for instance the ratings of a local television news program. Why do some stations have high ratings, and others have low ratings? What factors will shape a station's audience size in the future? Advertisers, as well as local station managers and programmers would probably have an interest in this sort of analysis. Imagine that you work for a station and you want to assess its situation.

A rating, of course, is a gross measure of audience size. Local news ratings, in particular, have an important impact on station profitability. To explain the size of a station's news audience, we should first consider structural factors. If audience size is to be expressed as an absolute number, we would need to know the size of the potential audience defined by the population in the market. At the same time, we would want to consider the nature of the station's coverage area. Is it a VHF or a UHF station? If it is the latter, you are probably already at a disadvantage. Is it carried on all the local cable systems? Is there anything about the station's signal or local geography that would limit the station's ability to reach all of the potential audience? Next, we would want to know the size of the audience at the time when the news is broadcast. An analysis of share data, of course, might overlook this, but since we are interested in ratings, the

A FRAMEWORK FOR ANALYSIS

bigger the available audience, the better are our chances of achieving a large rating. We might pay special attention to those segments of the audience that are more likely to be local news viewers. Experience tells us that these are probably going to be older adults. Next, we would consider a variety of program scheduling factors.

The first scheduling consideration would involve assessing the competition. Just how many competitors are there? As they increase in number, your ratings are likely to decrease. Do other stations enjoy any special advantages in covering the market? To what extent has cable television penetrated the market? What are your principle competitors likely to program opposite the news? Will you confront only news programs, or will the competition counter program with something different. The latter is much more likely if you are an affiliate with other independents in the market. If the available audience contains a large segment who are less likely to watch the news (e.g., children and young adults), that could damage your ratings. Consider the programming you have before and after your news. A highly rated lead-in is likely to help your ratings, especially if it attracts an audience of news viewers. If you are an affiliate, pay close attention to the strength of your network's news program. Research has shown that there is a very strong link between local and network news ratings.

More often than not, these structural factors explain most of the variations in local news ratings. A station can control some things, like lead-in programming. Other things, like the number of competitors it has, are beyond control. Because a single rating point might make a substantial difference in a station's profitability, however, consideration of individual factors may be warranted, especially if these are things a station can manipulate.

Among the most likely candidates for consideration are viewer preferences and awareness. Are there certain personalities, or program formats that are going to be more or less appealing to viewers? Every year, consultants to stations—*news doctors* they are called—charge large fees to make such determinations. Are there certain news stories that will better suit the needs and interests of local viewers? In markets that are not measured continuously, stations often schedule their most sensational and "sexiest" special reports to coincide with the ratings sweeps. A riveting investigative report is unlikely to boost a program's ratings, however, unless additional viewers are made aware of it. So stations must simultaneously engage in extraordinary promotional and advertising efforts. Of course, all the stations in a market are probably doing the same thing. Therefore, although catering to audience preferences is very important in principle, in practice, it may not make a huge ratings difference. Even so, a small edge can be crucial to a program's profitability. Just a few share points during the local news blocks can boost shares from "sign on to sign off," and, in the largest markets, that can mean millions of dollars in additional profits.

If we wanted to analyze some cumulative measure of audience behavior, the procedure would be much the same. Suppose, for example, we wanted to have a better sense of audience flow during prime time. This could have implications for programming, or the placement of advertising and promotional announcements. Like the preceding analysis of gross measures, we would recommend beginning the investigation with structural factors.

Essentially, you want to know the extent to which the audience for one program is duplicated in the audience of another program. Consider first the nature of the available audience. Ordinarily, if the two programs in question are scheduled back-to-back, the level of audience flow between them should be higher than if they are more widely spaced. That is because, on any given evening, the available audience in adjacent time periods is relatively constant. If total audience overlap is somehow abnormal, there could be a marked effect on program audience duplication. A network might also consider its coverage during each program. Often that would not be a factor, but if some affiliates have cleared one program and not the other, then the flow between them will obviously be impeded. Note the number of competitors available to the audience. The more there are, the more opportunities viewer have to defect, thereby reducing levels of duplication. In the break between adjacent programs, the number of competing programs that are just beginning should be inversely related to levels of duplication within a channel.

If those structural factors do not explain patterns of audience flow, begin to think through individual level determinants. Is the audience for one show likely to find the subsequent program appealing? Are competitors offering programming of a type that is going to draw heavily from another network's audience? If these assumptions about people's preferences are not explaining audience flow, what might be interfering with those preferences? Have program options been sufficiently well promoted? Are the programs in question scheduled at times when control of the set is likely to have shifted from one kind of viewer to another? Are group configurations, and hence the balance of power in decision making, changing? Any or all of these factors might explain variation in levels of duplication that are not accounted for by structural determinants.

The analysis of radio audience behavior follows along the same lines as television, except we typically define exposure in terms of stations and dayparts, rather than programs. Listeners do occasionally tune in to specific programs, but more often they will select a station without any notion of having chosen a discrete radio show. The key determinants of radio audience size and flow are still structural in nature, but the factors we have labeled *individual* may take on added salience. Radio stations usually operate in relatively competitive markets, and specialize in a particular kind of programming. The choice of stations is more likely to be the decision of a single individual than a group. Radio listeners are also likely to select a station by searching through a very limited repertoire, instead of consulting any sort of programming guide. For all these reasons, people's preferences and their awareness of what a particular station has to offer, weigh more heavily in the analysis of radio audience behavior.

In closing, we might comment briefly on the long-term nature of exposure to media. One danger of characterizing audience behavior as the result of nicely drawn arrows and boxes, is that things are made to seem simpler than they really are. For instance, the model defines exposure as the result, but not the cause, of other factors. Over a period of months, or even weeks, of course, ratings can have a substantial effect on the structure of the media. Programs are cancelled, new shows developed, schedules altered, and clearances changed, often on the basis of audience behavior. Such relationships have been the subject of a number of interesting investigations. Similarly, the model, as we have presented it, suggests a high degree of independence between audience and media factors. In the short term, that seems to be a workable assumption. Over the long haul, however, it could promote a distorted picture of audience behavior.

To address these issues, we have specified some long-term relationships between audience and media factors. For example, the growth of potential audiences and patterns of availability clearly affect the development of media services and programming strategies. Conversely, the structure and content of the media undoubtedly cultivate certain tastes, expectations, and habits on the part of the audience. These are important relationships, but not central to our purpose. Bearing such limitations in mind, we hope the model can provide a useful framework of evaluating ratings data, and exploiting the analytical techniques discussed in the remaining chapters.

RELATED READINGS

Barwise, P., & Ehrenberg, A. (1988). *Television and its audience.* London: Sage.
Bower, R. T. (1985). *The changing television audience in America.* New York: Columbia University Press.
Goodhardt, G. J., Ehrenberg, A. S. C., & Collins, M. A. (1987). *The television audience: Patterns of viewing* (2nd ed.). Aldershot, UK: Gower.
Heeter, C., & Greenberg, B. S. (1988). *Cable-viewing.* Norwood, NJ: Ablex.

Kubey, R., & Csikszentmihalyi, M. (1990). *Television and the quality of life: How viewing shapes everyday experience.* Hillsdale, NJ: Lawrence Erlbaum Associates.

McPhee, W. N. (1963). *Formal theories of mass behavior.* New York: The Free Press.

Owen, B. M., & Wildman, S. W. (1991). *Video economics.* Cambridge: Harvard University Press.

Rosengren, K. E., Wenner, L. A., & Palmgreen, P. (Eds.). (1985). *Media gratifications research: Current perspectives.* Beverly Hills: Sage.

Steiner, G. A. (1963). *The people look at television.* New York: Alfred A. Knopf.

Zillmann, D., & Bryant, J. (Eds.) (1985). *Selective exposure to communication.* Hillsdale, NJ: Lawrence Erlbaum Associates.

ANALYSIS OF GROSS MEASURES 8

Ratings data can be analyzed in many ways. In fact, the practice of ratings analysis may be constrained more by the skill and imagination of analysts than limitations inherent in the data. Describing common analytical techniques, as we do in chapters 8 and 9, runs the risk of discouraging inventive ways of looking at the data. That is certainly not our intent. In the preceding chapter we have suggested the kinds of research questions these data can address, described the nature of the data themselves, and developed a framework for understanding the audience behavior that the data represent. These insights, coupled with a knowledge of quantitative research methods, should be all that is needed to conduct ratings analyses.

There are, however, some advantages to becoming familiar with the most common techniques of ratings analysis. First, you do not have to "reinvent the wheel" every time you do an analysis of ratings data. Instead, you can use techniques that have been tested. Their strengths and limitations are, therefore, better known. Second, there is real value in some standardization of analytical techniques. As you see here, comparisons of one sort or another play an important part in ratings analysis. If everyone calculated the cost of reaching the audience in a different way, comparisons would be difficult or impossible to make. That would certainly limit the utility of the analysis. In the same vein, standardization can help us build a systematic body of knowledge about audiences and their role in the operation of the electronic media. If one study can be directly related to the next, progress and/or dead ends can be more readily identified.

Consistent with a distinction we made in chapter 7, we have organized analytical techniques into two chapters—those that deal with gross measures, and those dealing with cumulative measures. This distinction is not always easy to make. Analyses of one sort are often coupled with the other. In fact, there can be strict mathematical relationships between gross and cumulative measures. Nonetheless, we believe this scheme of organization will help the reader manage a potentially bewildering assortment of ways to manipulate the data.

Within each chapter, we go from the least complicated analytical techniques to the most complicated. Unfortunately, as we make this progression, our language becomes increasingly complex and arcane. The majority of analytical techniques described in each chapter require only an understanding of simple arithmetic. Some, however, involve the use of multivariate statistics. We try to keep the technical jargon to a minimum.

As we have noted, gross measures can be thought of as snapshots of the audience taken at a point in time. Included in this category are the measures themselves (e.g., ratings and shares), any subsequent manipulations of those measures (e.g., totaling GRPs), or analyses of the measures using additional data (e.g., calculating CPPs). Excluded from this category are audience measurements that require tracking individual audience members over time.

GROSS MEASURES

We begin this discussion by reviewing gross measures of the audience. Throughout the book, we have made frequent use of terms like *rating* and *share*. Although basic definitions of these terms were provided in the first chapter, those ignored a good many nuances that an analyst should know. In fact, it is important to recognize that these measures are themselves a kind of first-order data analysis. Ratings, shares, and gross audience projections are all the result of mathematical operations being applied to the database.

Projected audiences are the most basic gross measurements of the audience. They are estimates of absolute audience size, intended to answer the question "how many people watched or listened . . .?" Audience projections can be made for specific programs, specific stations, or for all those using a medium at any point in time. Projections can be made for households, persons, or various subsets of the audience (e.g., how many men 18 to 49 watched the news). Most of the numbers reported in a ratings book are simply estimates of absolute audience size.

Projections of this sort are necessarily made on the basis of sample information. The most straightforward method of projection is to determine the proportion of the sample using a program, station, or medium, and multiply that by the size of the population. For example, if we wanted to know how many households watched Program Z, would look at the sample, note that 20% watched Z, and multiply that by the estimated number of TV households in the market, say 100,000. The projected number of TVHH watching Program Z would therefore be 20,000. That proportion is, of course, a rating. Hence, projected audiences can be derived by the following equation:

$$\text{Rating (\%)} \times \text{Population} = \text{Projected Audience}$$

This is the approach that Nielsen uses with all its metered samples, both nationwide and in large markets. It assumes that the in-tab sample, without further adjustments, adequately represents the population. Many local market samples, however, do not meet that assumption. You will recall, for example, that in-tab diary samples tend to over-represent some groups and under-represent others. In such instances, it is common to weight the responses of under-represented groups more heavily than others. The specific variables used for purposes of weighting, and the way these weights are combined varies from market to market, and ratings company to ratings company. The end result, however derived, is that the weighted responses of households or individuals, expressed as households per diary value (HPDV) or persons per diary value (PPDV), are combined to project audience size. Unlike the simple procedure described above here projected audiences must be determined before ratings. In fact, if sample weighting or balancing is used, audience projections are actually used to calculate a rating, not vice versa.

We saw, in chapter 6, that audience projections for radio tell you how many hundreds of people listened to a station in an average quarter hour. The total number of people listening to radio without regard to stations, is called persons using radio, or (PUR). In television, audiences are typically associated with specific programs in specific quarter hours. Depending on the unit of analysis, the total size of the TV audience is called households using television, (HUT), persons using television (PUT), or persons viewing television (PVT). All these numbers express the absolute size of the audience at a single or average point in time.

Audience projections, used in the context of advertising, will sometimes be added to produce a number called gross audience or *gross impressions*. This is a summation of program or station audiences across different points in time. Those points in time are usually defined by an advertiser's schedule of spots. Table 8.1 is a simple example of how gross impressions, for women 18–49, would be determined for a commercial message that aired at four different times.

Gross impressions are just like GRPs, except they are expressed as a whole number. They provide a crude measure of the total weight of audi-

TABLE 8.1
Determining Gross Impressions

Spot Availability	Audience of Women 18–49
Monday, 10 a.m.	2,500
Wednesday, 11 a.m.	2,000
Thursday, 4 p.m.	3,500
Friday, 9 p.m.	1,500
Total (Gross Impressions)	9,500

ence exposure to a particular message or campaign. They do not take frequency of exposure or audience duplication into account. As a result, 10,000 gross impressions might mean that 10,000 people saw a message once, or 1,000 people saw it 10 times.

Ratings are the most familiar of all gross measures of the audience. Unlike projected audience, they express the size of the audience as a percentage of the total population, rather than a whole number. The simplest calculation for a rating, therefore, is to divide a station or program audience by the total potential audience. In practice, the "%" is understood, so a program with 30% of the audience is said to have a rating of 30.

The potential audience on which a rating is based can vary. Household ratings for the broadcast network are based on all U.S. households equipped with television (TVHH). But ratings can also be based on people, or different categories of people. For example, a rating service will often report ratings for men 18–34, women 25–49, and so on. Local market reports will have station ratings for different market areas, like ADI and metro ratings. Some national cable networks will base their ratings not on all TVHH, or even all cable households, but only on those homes that can receive the network's programming. Although there is a certain rationale to that, such variation can affect our interpretation of the data. A ratings analyst should, therefore, be aware of the potential audience on which a rating is based.

In addition to these distinctions, are several different kinds of ratings calculations. These are summarized in Table 8.2. To simplify the wording in the table, everything is described in terms of television, with TV households (TVHH) as the unit of analysis. Radio ratings, and TV ratings using persons as the unit of analysis, would be just the same, except they would use slightly different terminology (e.g., PUR vs. HUT).

Average quarter-hour ratings are used in both radio and television. They are a convenient way to represent the average size of an audience over some specified period of time. Usually, that time period is a standard daypart like Monday through Friday 6 a.m. to 10 p.m. (or M–F 6–10).

ANALYSIS OF GROSS MEASURES

TABLE 8.2
Ratings Computations[a]

Basic rating (R)	$R\ (\%) =$	$\dfrac{\text{TVHH watching program or station}}{\text{Total TVHH}}$
Quarter-hour rating (QH)	$QHR =$	$\dfrac{\text{TVHH watching more that 5 minutes in a quarter hour}}{\text{Total TVHH}}$
Average quarter-hour rating (AQH)	$AQH =$	$\dfrac{\text{Sum of quarter hour ratings}}{\text{Number of quarter hours}}$
Average audience rating[b] (AA)	$AA =$	$\dfrac{\text{Total minutes all TVHH spend watching a program}}{\text{Program duration in minutes X total TVHH}}$
Total Audience Rating (TA)	$TA =$	$\dfrac{\text{TVHH watching program for more than 5 minutes}}{\text{Total TVHH}}$
HUT rating (HR)	$HR =$	$\dfrac{\text{Projected HUT level}}{\text{Total TVHH}}$
Gross rating points (GRP)	$GRP =$	$R_1 + R_2 + R_3 + R_4 + R_5 \ldots + R_n$

[a] The precise method for computing a ratings depends on whether the responses of sample members are differentially weighted. When they are, program audiences must be projected and then divided by the total estimated population. When the responses of sample members are not weighted, or have equal weights, proportions within the sample itself determine the ratings and subsequent audience projections.

[b] In this computation, the number of minutes each TVHH spends watching a program is totaled across all TVHH. This is divided by the total possible number of minutes that could have been watched, as determined by multiplying program duration in minutes by total TVHH. AA can also be reported for specific quarter hours within the program, in which case the denominator is 15 × total TVHH.

Although this is often referred to as *morning drive* time, as you note later, the majority of listeners are at home, listening, while they get ready to leave the house. Within a daypart, of course, audience levels may fluctuate from one quarter hour to the next. In television, advertisers who are buying time on a particular program might, understandably, want more precise information on audience size and composition. Television ratings books will, therefore, report program audiences in specific quarter hours. Similarly, radio ratings are also presented hour-by-hour, and can be analyzed for each quarter hour if desired.

The most narrowly defined audience rating is the average audience rating reported in the NTI. This rating expresses the size of the audience in an average minute, within a given quarter hour. To obtain that level of precision, metering devices must be used. As a result, this sort of rating cannot be reported for diary-based data.

Also summarized in Table 8.2 are GRPs and HUT ratings. These are analogous to gross impressions and HUTs, respectively. They carry essentially the same information as those projections of audience size; they are simply expressed as percentages instead of whole numbers. They are also subject to the same interpretive limitations as their counterparts. We should point out that, strictly speaking, reporting HUT or PUT as percentages means they are a kind of rating. To avoid confusion, we will refer to them as such. In practice, however, these percentages are usually called HUTs or PUTs, without appending the word "rating."

Shares are the third major measure of audience size. They express audience size as a percentage of those using the medium at a point in time. The basic equation for determining audience share among TV households is:

$$\frac{\text{\# of TVHH Tuned to Station or Program}}{\text{HUT}} = \text{Share}$$

The calculation of person shares is exactly the same, except persons and PUT levels are in the numerator and denominator, respectively. In either case, the rating and share of a given program or station have the same number in the numerator. The difference is in the denominator. Because HUT or PUT levels are always less than the total potential audience, a program's share will always be larger than its rating.

Like ratings, audience shares can be determined for various subsets of the total audience. Unlike ratings, however, shares are of somewhat limited value in buying and selling audiences. Although shares indicate performance relative to the competition, they do not convey information about actual size of the audience, and that is what advertisers are most often interested in. The only way that a share can reveal information about total audience size is when it is related to its associated HUT level, as follows:

$$\text{Program Share} \times \text{HUT} = \text{Projected Program Audience}$$

or

$$\text{Program Share} \times \text{HUT Rating} = \text{Program Rating}$$

Audience shares can also be calculated over periods of time longer than program lengths. In ratings books, for example, audience shares are often reported for entire dayparts. When long-term average share calculations are made, the preferred method is to derive the average quarter hour

ANALYSIS OF GROSS MEASURES

(AQH) share from the average quarter hour rating within the same daypart. The following equation summarizes how such a daypart share might be calculated with TV data:

$$\frac{\text{AQH Rating}}{\text{AQH HUT Rating}} = \text{AQH Share}$$

Unlike AQH ratings, it is not appropriate to calculate AQH shares by adding a station's audience share in each quarter hour, and dividing by the number of quarter hours. That is because each audience share has a different denominator, and it would distort the average to give them equal weight.

Defining audience size in these ways presents some interesting problems. Many of them occur when households are the unit of analysis. We noted in chapter 7, that most homes have more than one television set. Suppose that a household is watching two different programs on different sets. To which station should that home be attributed? Standard practice in the ratings business is to credit the household to both station's audiences. In other words, it counts in the calculation of each station's household rating and share. However, it will only be allowed to count once in the calculation of HUT levels. This means that, contrary to what most textbooks tell you, the sum of all program ratings can exceed the HUT rating, and the sum of all program shares can exceed 100. This was no problem in the early days of TV, when most of these methods evolved, because most homes had only one TV. But now more than two thirds of all homes have multiple sets. There is an average of just about two TV sets per home. Further, it is now possible for a household to be watching several programs while simultaneously taping other programs on videotape for later viewing. Because of multiple sets, mostly individual listening, and much away-from-home listening radio measurement is based on "persons" rather than "households."

Because households are typically collections of two or more people, household ratings tend to be higher than person ratings. Imagine, for example, that some market has 100 homes, with four people living in each. Suppose that one person in each household was watching Station Z. That would mean that Station Z had a TVHH rating of 100 and a person rating of 25. Some programs, like "family shows," do better at attracting groups of viewers, whereas others garner more solitary viewing. It is, therefore, worth keeping an eye on discrepancies between household and person ratings, because differences between the two can be substantial.

Even when people are the unit of analysis, aberrations in audience size can occur. Most ratings services require that a person be in a program, or quarter hour, audience for at least 5 minutes in order to be counted. That means it is quite possible for a person to be in two programs in a quarter hour, or to show up in several program audiences of longer duration. This creates a problem analogous to multiple set use at the household level. In the old days, when person ratings could only be made with diaries, and the audience had to get up to change the channel, it was not much of a problem. Today, with peoplemeters tracking a population that has remote control devices (in 75% of homes), and dozens of channels from which to choose, the potential for viewers to show up in more than one program audience is considerably enhanced.

The most common way to extend the data reported in a typical ratings book is to introduce information on the cost on reaching the audience. Cost calculations are an important tool for those who must buy and sell media audiences. There are two such calculations in wide use. Both are based on manipulations of gross audience measurements.

Cost per thousand (CPM), as the name implies, tells you how much it costs to reach 1,000 members of a target audience. It is a yardstick that can be used to compare stations or networks with different audiences and different rate structures. The standard formula for computing CPMs is:

$$\frac{\text{Cost of Spot (\$'s)} \times 1000}{\text{Projected Target Audience}} = \text{CPM}$$

The projected target audience is expressed as a whole number. It could simply be the number of households delivered by the spot in question, or it could also be men 18–49, working women, teens 12–17, and so on. CPMs can be calculated for whatever audience is most relevant to the advertiser, as long as the ratings data can be calculated to project that audience. Occasionally, when a large number of spots are running, it is more convenient to compute the average CPM for the schedule in the following way:

$$\frac{\text{Cost of Schedule (\$'s)} \times 1000}{\text{Target Gross Impressions}} = \text{Average CPM}$$

CPMs are the most widely used measure of the advertising media's cost efficiency. They can be calculated to gauge relative costs within a medium, or used to make comparisons across different media. In print, for example, the cost of a black-and-white page or a newspaper's line rate, is divided by its circulation or the number of readers it delivers. In chapter 3 we presented CPM trends across radio, television, and print media. Comparisons within a medium, are generally easier to interpret than intermedia comparisons. As long as target audiences are defined in the same way, CPMs do a good job of revealing which spot is more cost efficient. There is less agreement on what is the magazine equivalent of a 30-second spot.

The electronic media have a unique form of cost calculation called cost per point (CPP). Like CPM, it is a yardstick for making cost efficiency

ANALYSIS OF GROSS MEASURES

comparisons, except here the unit of measurement is not 1,000s of audience members, but ratings points. CPP is computed as follows:

$$\frac{\text{Cost of Spot (\$)}}{\text{Target Audience Rating}} = \text{CPP}$$

An alternative method for calculating CPP can be used when a large number of spots are being run and an average CPP is of more interest that the efficiency of any one commercial. This is sometimes called the cost per gross ratings point.

$$\frac{\text{Cost of Schedule (\$)}}{\text{Gross Rating Points}} = \text{CPGRP}$$

As you know by now, there are different kinds of ratings. In network television, average audience ratings (AA) are preferred for CPP computations, because they more accurately express the size of the audience at the moment a spot is run. For ratings based on diary data, a quarter-hour rating is used. In radio it must be an average quarter hour. In television a specific quarter hour is preferable. Television market reports also estimate "station break" ratings by averaging quarter hours before and after the break. If that is when the spot has run, that is the most appropriate rating to use.

CPP measures are part of the everyday language of people who specialize in broadcast advertising. Stations representatives, and the media buyers with whom they deal, often conduct their negotiations on a CPP basis. This measure of cost efficiency has the additional advantage relating directly to GRPs, which are commonly used to define the size of an advertising campaign. CPPs, however, have two limiting characteristics that affect their use and interpretation.

First, they are simply less precise than CPMs. Ratings points are rarely carried beyond one decimal place. They must, therefore, be rounded. Rounding off network audiences in this way can add or subtract tens of thousands of people from the audience, causing an unnecessary reduction in the accuracy of cost calculations. Second, ratings are based on different potential audiences. We would expect the CPP in New York to be more than it is in Louisville, because each point represents many more people. But how many more? CPMs would be much more interpretable. Even within a market, problems can crop up. Radio stations, whose signals may cover only part of a market, should be especially alert to CPP buying criteria. It is quite possible, for example, that one station delivers most of its audience within the metro area, whereas another has an audience of equal size located mostly outside the metro. If CPPs in the market are based on metro ratings, the second station could be at an unfair, and unnecessary, disadvantage.

COMPARISONS

Comparing the gross measures we have just reviewed is the most common form of ratings analysis. There are an endless number of comparisons that can be made. They might be designed to show the superiority of one station over another, the relative cost efficiency of different advertising media, or the success of one program format as opposed to another. There are certainly more of these than we can catalog in this chapter. We can, however, provide illustrative examples of comparisons that may be useful in buying or selling time, programming, or simply reaching a better understanding of the electronic media and their audience.

One area of comparison, deals with the size and composition of the available audience. You know from chapter 7, that the nature of the available audience is a powerful determinant of station or program audiences. An analyst might, therefore, want to begin by taking a closer look at who is watching or listening at different times. This kind of analysis could certainly be of interest to a programmer who must be cognizant of the ebb and flow of different audience segments when deciding what programming to run. It might also be of value to an advertiser or media buyer, who wants to know when a certain kind of audience is most available. The most straightforward method of comparison is to graph the size of various audiences segments at different hours throughout the day. So at this point we return briefly to the concept of available audiences.

The single most important factor effecting the size of broadcast audiences is when people are available to listen. Work hours, school and transportation schedules, meal times, and the seasons—especially the length of the day and warm weather when people are much more likely to be outdoors—are the strongest influence on when people are available, and interested in using mass media. There are no regular surveys that provide detailed information on such availabilities. However, based on several older studies we have reconstructed, Fig. 8.1A is intended to give the reader a rough idea of the availability of men, women, teens, and children throughout the day.

Holidays, special events, and coverage of especially important news stories can certainly alter these patterns of availability, but as a rule, they translate rather directly into the patterns of media use depicted in Fig. 8.1B. The instances when a single program, or big event, has influenced a rise in HUTs are so rare as to be counted on only one or two hands—at least since the very earliest days of TV. The assassination of President Kennedy, and attempted assassination of President Reagan are the most profound, if morbid examples. The incredible success of "The Cosby Show" beginning in September 1984 demonstrated that one program could raise HUTs on one night. Maybe the most unusual example

ANALYSIS OF GROSS MEASURES

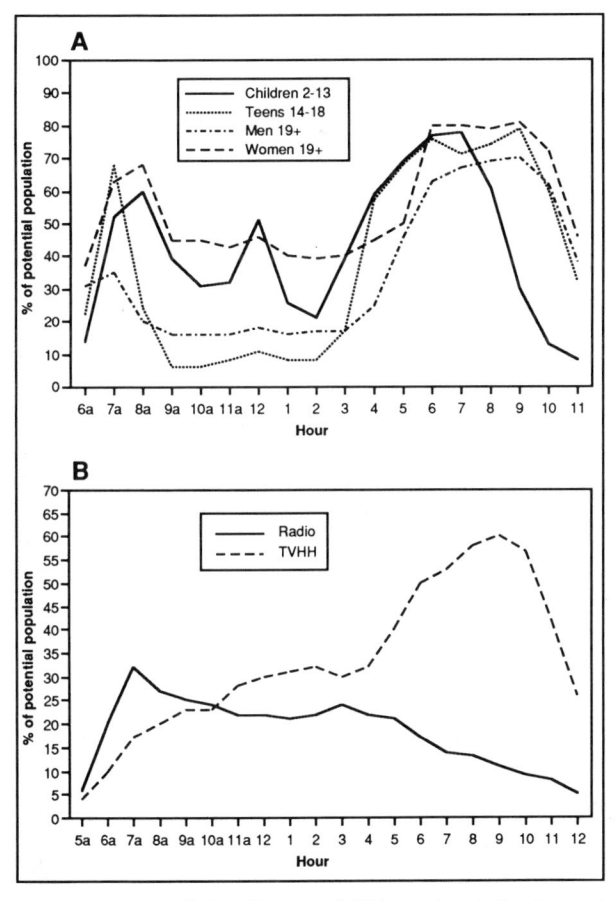

FIG. 8.1. Hypothetical audience availabilities and typical patterns of radio and television use.

was a temporary increase in overnight viewing—and home taping—in January 1987 during the ESPN live coverage of the America's Cup from Australia.

With some exceptions, however, the best indication of how many people can and will use the media is indicated by the reports of when they do so. Any new program, or even new cable network, that plans to find an audience among those who are not already listening or viewing is very, very unlikely to be successful. New programs, formats, and program services for the most part divide the existing potential and available audiences in to smaller "pieces of the pie" rather than cause "new viewers" to tune in. The most obvious evidence of this is the recent decline in national network share.

Thus, to plot when various audience segments are using a medium in a market is a valuable starting point for the audience analyst. Radio market reports will have a section of "Hour By Hour" audience estimates with people using radio (PUR) levels, as well as different station audiences. Television reports will typically estimate audiences by the quarter hour. For illustration purposes, Fig. 8.2 shows radio listening—at home, at work, and in cars—for three different gender-age categories.

The data paint rather different pictures of radio use for each demographic group. Note especially the amount of radio use by working men on the job and in their vehicles. For teens, while school is in session, there is very little daytime listening but a lot at night. Older people are heavy radio users mostly at home. The distinct patterns of radio use revealed by this comparison clarifies the parameters within which the programmer must operate.

Advertisers, of course, must eventually commit to buying time on specific stations or networks. To do so, they need to determine the most effective way to reach their target audience. This relatively simple requirement can trigger a torrent of ratings comparisons. From the time buyer's perspective, the comparisons should be responsive to the advertiser's need to reach a certain kind of audience in a cost efficient manner. From the time seller's perspective, the comparison's should also show his or her audiences in the best possible light. Although these two objectives are not mutually exclusive, they can cause audience analysts to look at ratings data in different ways.

The simplest form of ratings analysis is compare station or program audiences in terms of their size. This can be determined by ranking each program, station, or network by its average rating and, in effect, declaring a winner. One need only glance at the trade press to get a sense of how important it is to be "Number 1" by some measure. Of course, its difficult for everyone to be Number 1. Further, buying time on the top-rated station may not be the most effective way to spend an advertising budget. So, comparisons of sheer audience size are typically qualified by some consideration of audience composition.

The relevant definition of audience composition is usually determined by an advertiser. An "avail request," for instance, will usually specify the target audience in demographics. If the advertiser has a primary audience of women ages 18–24, it would make sense for the analyst to rank programs, not by total audience size, but by ratings among women 18–24. In all probability this would produce a different rank ordering of programs, and perhaps even a different Number 1. For radio stations, which often specialize in a certain demographic, ranking within audience subsets can allow several stations to claim they are Number 1. Table 8.3 is a summary of the audience breakdowns that are usually reported in ratings books. In

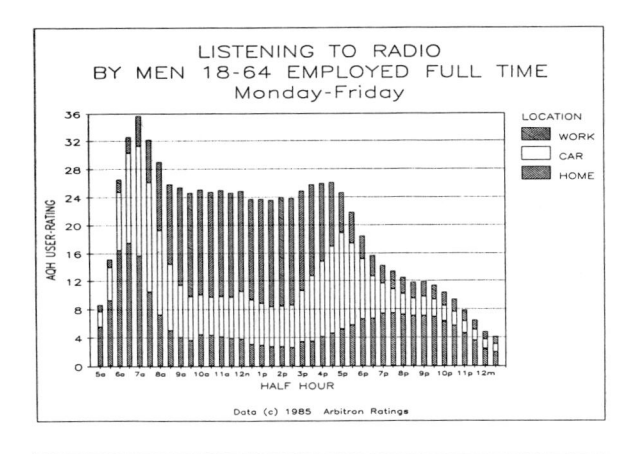

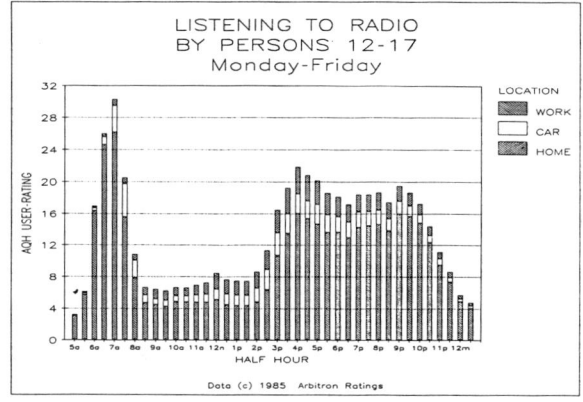

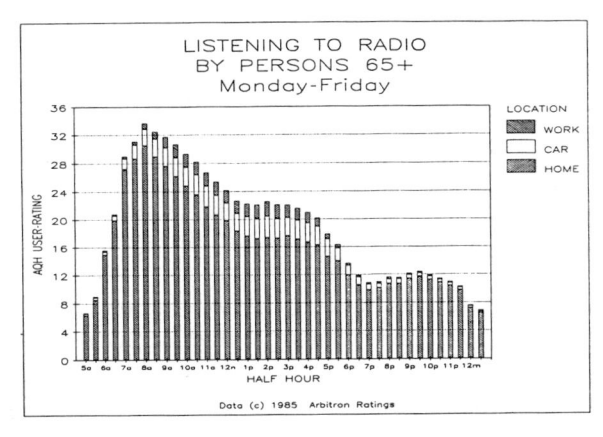

FIG. 8.2. Comparison of radio use by location within demographic groups (Based on data provided by the Arbitron Company. Reprinted with permission of Audience Research Analysis).

TABLE 8.3
Common Demographic Breakdowns

Women	Men	Persons
Ages		
—	—	2+[a]
—	—	2–11[a]
—	—	6–11[a]
—	—	12+
12–17	—	12–17
12–24	12–24	12–24
18+[b]	18+	18+
18–24	18–24	—
18–34	18–34	18–34
18–49[c]	18–49	18–49
25–34	25–34	—
25–49	25–49	25–49
25–54	25–54	25–54
35+	35+	35+
35–44	35–44	—
35–64	35–64	—
45–54	45–54	—
55+	55+	55+
55–64	55–64	—
65+	65+	—

[a] Television only, radio ratings do not include those under 12.
[b] A subset of women 18+ who work outside the home for 30 or more hours a week are also commonly reported in television market reports.
[c] A subset of women 19–49 who are the "Lady of House" and have children under 3 are reported by NTI as "LOH 18–49 W/CH <3".

principle, any one of these audience subsets could produce a different ratings winner.

At this point, we should emphasize a problem in ratings analysis, about which researchers are quite sensitive. Any time you begin to analyze or compare subsets of the audience, it reduces the actual sample size on which those comparisons are based. It is easy for casual users to ignore or forget this because published ratings seem so authoritative once the ink on the page has dried. But remember that ratings estimates are subject to sampling error, and the amount of error increases as the sample size decreases. That means, for instance, that the difference between being Number 1 and Number 2 among men ages 18–24 might be a chance occurrence, rather than a real difference. A researcher would say the difference was not "statistically significant." The same phenomenon produces what people in the industry call *bounce*. It is a change in station ratings from one book to the next, that is a result of a sampling error,

ANALYSIS OF GROSS MEASURES

rather than any real change in the station's audience size. An analyst should never be impressed with small differences, especially if they are based on small samples.

Having so cautioned, we must also point out that the business of making comparisons can be, and is, done using things other than shear audience size. Ratings data can be adjusted in a way that highlights audience composition, and then ranked. This may produce very different rank orderings. There are two techniques sometimes used to make these adjustments.

Indexing is a common way to make comparisons across scores. An index number simply expresses an individual score, like a rating or CPM, relative to some standard or base value. The basic formula for creating index numbers is as follows:

$$\frac{\text{Score} \times 100}{\text{Base Value}} = \text{Index Number}$$

Usually the base value is fixed at a point in time to give the analyst an indication of how some variable is changing. Current CPMs, for instance, are often indexed to their levels in an earlier year. Base values have been determined in other ways, however. Suppose a program had a high ratings among women 18–24, but a low rating overall. An index number could be created by using the overall rating as a base value. That would make the target audience rating look strong by comparison. CPM index numbers have also been created by comparing individual market CPMs to an average CPM across markets (Poltrack, 1983).

Another way to represent the same data is to compute a percentage sometimes labeled *target audience efficiency* (TAE). This expresses a station's target audience, measured in whole numbers, as some portion of its total audience. With this sort of analysis, it might be possible to argue that although a station's audience is not big in absolute terms, a relatively high percentage is in the advertiser's target, hence buying that audience is "efficient." In radio, the computational formula would be as follows:

$$\frac{\text{Station's AQH Persons (Target)}}{\text{Station's 12+ AQH Persons}} = \text{TAE}$$

With the appropriate substitutions, this could also be done for specific television programs. The results could then be rank ordered to represent a particular program or average quarter hour as the most efficient vehicle for reaching the target audience. Whether an advertiser finds this to be a persuasive argument, however, is another matter.

Thus far, we have defined target audiences only in terms of two demographic variables; age and gender. These are the segmentation variables most commonly used to specify an advertiser's audience objectives. Age and gender, of course, may not be the most relevant descriptors of an advertiser's target market. Income, buying habits, lifestyle, and a host of other variables might be of critical importance to someone buying advertising time. If the seller of time could define target audiences in those terms, it might be an effective sales tool. Unfortunately, ratings books report very little of that kind of specialized information.

We noted in chapter 6, however, that ratings services are capable of producing "customized" ratings reports. In fact, with the widespread use of personal computers and telephone access to databanks, this sort of customization is becoming increasingly common. As a consequence, it is now possible to describe audiences in ways that are not reported on the printed page of a ratings book.

It is common, these days, for the ratings services to keep track of the zip code in which each member of the sample lives. Zip code information is valued because it is thought that knowing where a person lives can reveal a great deal about that individual. For example, inferences can be made about household incomes, occupations, ethnicity, education levels, lifestyles, and so on. As long as sample sizes are sufficiently large, these inferences will, on average, be reasonably accurate. In fact, some companies specialize in analyzing zip code areas and grouping those with similar characteristics. The use of *geodemography,* as it is sometimes called, means that audiences can be defined and compared in a virtually unlimited number of ways.

Table 8.4 combines many of the comparison techniques we have described so far. It was produced by a computer program marketed by Arbitron. This program requires the analyst to tap into a special database containing information on everything from people's buying habits to their political beliefs.

This particular run was done to identify programs with the largest audiences of 18- to 49-year-olds, whose political outlooks were somewhat to very conservative. Programs were ranked according to their ratings among this targeted subset of the audience. In this particular sweep, "NFL Monday Night Football" ranked highest on that criterion. As it turned out, it also ranked highest in the overall ADI ratings. The Number 2 program, however, had a high target rating, but a relatively low ADI rating overall (i.e., 11.4 vs. 6.0). The computer program also printed out a "target rating index" which relates each target rating to the show's ADI rating. In the second to last column, the print out lists "target % of ADI audience," which is the same as a TAE calculation.

Of course, none of these audience comparisons will necessarily convince an advertiser to buy time. As with any product for sale, no matter how useful or nicely packaged, the question usually comes down to how much

ANALYSIS OF GROSS MEASURES

TABLE 8.4
Customized Audience Breakdown in Arbitron Product Target AID[a]

```
ARBITRON PRODUCT TARGET AID / 04-10-1988              1      ADVERTISER:
                                                             AGENCY:
TARGET VIEWER ANALYSIS REPORT: SACRAMENTO-STOCKTON (West)    CONTACT:
                                                             ADDRESS:
           SIMMONS PRODUCT CATEGORY: 501 POLITICAL OUTLOOK IS VERY/SOMEWHAT CONSERVATIVE
        DONNELLEY ClusterPlus SELECTIONS: 9 29 32 33 41
              ARBITRON SURVEY MONTH: 11/1987                 DEMOGRAPHIC CATEGORY: P 18-49
                         SORTED BY: Target Rating           DAYPART SET: NOV 87
                                                            NUMBER OF DAYPARTS: 33
    CLUSTER POPULATION:      162,434  CLUSTER IN-TAB:     182
         ADI POPULATION:   1,220,560  TOTAL IN-TAB:     1,383
                                                                                           TARGET   TARGET %
                                                      ADI       ADI      ******            RATING  OF ADI
                                                      ADI       ADI      TARGET   TARGET   INDEX   AUDIENCE  RNK
RNK STATION  DAYPART/PROGRAM         DAY(S)    TIMES      WEEK RATING   AUDIENCE  RATING  AUDIENCE
  1  KOVR   NFL MONDAY FOOTBALL     MONDAY   6:00P- 9:30P 1234  12.2   150,000    14.1    23,000    116       15   1
  2  KCRA   CHANNEL 3 REPORTS       MON-FRI  5:00P- 6:00P 1234   6.0    74,000    11.4    19,000    190       25   2
  3  KCRA   NBC PRIME TIME          MON-FRI  8:00P-11:00P 1234  10.6   130,000    11.2    18,000    106       14   3
  4  KOVR   ABC PRIME TIME          TUE-FRI  8:00P-11:00P 1234   7.9    98,000     8.6    14,000    109       14   4
  5  KXTV   CBS PRIME TIME          MON-FRI  8:00P-11:00P 1234   7.4    91,000     7.4    12,000    100       13   5
  6  KXTV   WHEEL FORTUNE/JEOPDY    MON-FRI  7:00P- 8:00P 1234   6.5    80,000     6.5    11,000    100       13   6
  7  KCRA   CHANNEL 3 REPORTS       MON-FRI 11:00P-11:30P 1234   4.8    59,000     6.0    10,000    125       17   7
  8  KCRA   CHANNEL 3 SUNRISE       MON-FRI  6:00A- 7:00A 1234   3.3    40,000     5.7     9,000    173       23   8
  9  KTXL   HILL STREET BLUES       MON-FRI  6:00P- 7:00P 1234   3.9    48,000     5.0     8,000    128       17   9
 10  KCRA   DONAHUE                 MON-FRI  4:00P- 5:00P 1234   3.9    48,000     5.0     8,000    128       17  10
 11  KCRA   WIL SHRINER             MON-FRI  7:00P- 8:00P 1234   2.8    35,000     4.8     8,000    171       22  11
 12  KCRA   CHANNEL 3 REPORTS       MON-FRI  6:30P- 7:00P 1234   4.7    58,000     4.7     8,000    100       13  12
 13  KOVR   AFTERNOON ROTATION      MON-FRI  1:00P- 3:00P 1234   3.9    48,000     4.3     7,000    110       15  13
 14  KTXL   BIG MOVIE               MON-FRI  8:00P-10:00P 1234   4.7    58,000     4.2     7,000     89       12  14
 15  KOVR   SQUARES/WIN,LOSE,DRW    TUE-FRI  7:00P- 8:00P 1234   4.0    49,000     4.2     7,000    105       14  15
 16  KRBK   WORLD DISNEY            MON-FRI  7:00P- 8:00P 1234   3.5    43,000     4.1     7,000    117       16  16
 17  KCRA   TONIGHT SHOW            MON-FRI 11:30P-12:30A 1234   2.7    33,000     4.0     7,000    148       20  17
 18  KTXL   FAMILY TIES/CHEERS      MON-FRI  7:00P- 8:00P 1234   5.7    70,000     3.4     6,000     60        8  18
 19  KCRA   TODAY SHOW              MON-FRI  7:00A- 9:00A 1234   2.6    32,000     3.2     5,000    123       16  19
 20  KCRA   DAYS OF OUR LIVES       MON-FRI  3:00P- 4:00P 1234   5.6    68,000     3.1     5,000     55        7  20
 21  KOVR   KOVR 13 NEWS            MON-FRI 11:00P-11:30P 1234   2.2    27,000     3.1     5,000    141       19  21
 22  KOVR   KOVR 13 NEWS            MON-FRI  5:00P- 6:00P 1234   2.5    31,000     2.9     5,000    116       16  22
 23  KOVR   KOVR 13 NEWS            MON-FRI 12:00N- 1:00P 1234   1.8    23,000     2.9     5,000    161       21  23
 24  KXTV   OPRAH WINFREY           MON-FRI  4:00P- 5:00P 1234   4.0    49,000     2.9     5,000     73       10  24
 25  KOVR   MORNING ROTATION        MON-FRI  9:00A-12:00N 1234   2.5    31,000     2.9     5,000    116       15  25
 26  KOVR   LOVE CONNECTION         TUE-FRI  6:30P- 7:00P 1234   3.4    42,000     2.9     5,000     85       11  26
 27  KXTV   NEWS 10                 MON-FRI  5:00P- 6:00P 1234   3.5    43,000     2.7     5,000     77       10  27
 28  KXTV   MORNING ROTATION        MON-FRI  9:00A-12:00N 1234   1.8    23,000     2.7     5,000    150       20  28
 29  KXTV   ENTERTAINMENT TONITE    MON-FRI  6:30P- 7:00P 1234   2.9    36,000     2.6     4,000     90       12  29
 30  KRBK   8PM MOVIE               MON-FRI  8:00P-10:00P 1234   2.2    27,000     2.4     4,000    109       14  30
 31  KCRA   AFTERNOON ROTATION      MON-FRI  1:00P- 3:00P 1234   1.8    23,000     2.3     4,000    128       17  31
 32  KRBK   THREES COMPANY          MON-FRI  6:00P- 6:30P 1234   3.6    44,000     2.3     4,000     64        9  32
 33  KTXL   10 O'CLOCK NEWS         MON-FRI 10:00P-11:00P 1234   1.9    24,000     2.3     4,000    121       16  33
 34  KRBK   SPOONS/HAPPY DAYS       MON-FRI  5:00P- 6:00P 1234   2.1    26,000     2.2     4,000    105       14  34
 35  KRBK   TOO CLOSE COMFORT       MON-FRI  6:30P- 7:00P 1234   2.9    36,000     2.0     3,000     69        9  35
 36  KOVR   ABC MORNING NEWS        MON-FRI  6:00A- 7:00A 1234   0.7     9,000     2.0     3,000    286       37  36
 37  KCRA   CHANNEL 3 NOON          MON-FRI 12:00N- 1:00P 1234   1.7    22,000     2.0     3,000    118       15  37
 38  KOVR   NIGHTLINE               MON-FRI 11:30P-12:00M 1234   1.3    17,000     1.6     3,000    123       16  38
 39  KTXL   FACTS LIFE/WKRP         MON-FRI  5:00P- 6:00P 1234   1.8    22,000     1.6     3,000     89       12  39
 40  KRBK   BOB NEWHART             MON-FRI 10:30P-11:00P 1234   0.7     9,000     1.4     2,000    200       27  40
 41  KTXL   AFTERNOON ROTATION      MON-FRI 12:00N- 2:30P 1234   1.1    14,000     1.4     2,000    127       17  41

Copyright 1986,1987 SIMMONS MARKET RESEARCH BUREAU
Copyright 1987,1988 DONNELLEY MARKETING INFORMATION SERVICES
Copyright 1988 ARBITRON RATINGS COMPANY
```

[a] *Note:* Reprinted with permission of the Arbitron Company

it costs. In this context, CPM and CPP comparisons are critical. Such comparisons might be designed to illuminate the efficiency of buying one program, station, or daypart as opposed to another. Table 8.5 compares primetime CPMs for network and spot television. It also reports the CPM's as indexed relative to 1976 prices. In this case, the index numbers suggest that although CPMs are lower on the networks, they have increased at a faster rate than spot CPM's.

We pointed out than audience ratings usually have more sales applications that audience shares. Share data can, nevertheless, be useful in promoting a particular station, program, or an entire medium. Even though audience shares do not always total to 100, most people familiar with the concept of market shares will expect them to. It is quite common, and often very effective, to represent audience shares in the form of a pie

TABLE 8.5
CPM Trends in Prime Time Television[a]

	Network		Spot	
	CPM TVHH ($)	Index	CPM TVHH ($)	Index
1976	3.49	100	4.86	100
1977	4.13	118	5.90	121
1978	4.41	126	6.42	132
1979	5.06	145	6.88	142
1980	4.94	142	7.35	151
1981	5.84	167	8.15	168
1982	6.44	185	8.69	179
1983	6.94	199	8.82	181
1984	8.08	232	8.89	183
1985	8.06	231	10.10	208
1986	8.78	252	10.29	211
1987	10.13	290	10.01	206
1988	9.46	271	11.58	238

[a] *Source:* Burnett (1989).

chart. Figure 8.3, for instance, is a series of pie charts prepared by the Cable Advertising Bureau (CAB), to dramatize the share of that cable services claim among various household types.

Rating and share comparisons can also be quite useful to programmers. Consider, again, how zip code areas can be used to segment and compare station audiences. A radio station might compare its ratings across geographic areas within the market. Because different formats tend to appeal to different kinds of people, a programmer who knows the market should have some sense of where his or her listeners are likely to live. If a station places a strong signal over an area with the kind of population that should like its format, but has few listeners, special promotions might be called for. For example, one station that made such a discovery decided to place

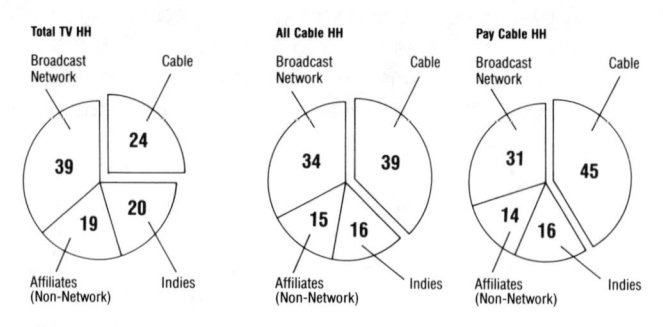

FIG. 8.3. Audience shares as pie charts (reprinted with permission of Cabletelevision Advertising Bureau).

ANALYSIS OF GROSS MEASURES

outdoor advertising and conduct a series of remote broadcasts in the areas where it was underperforming.

Radio programmers may also find it useful to represent the audience for each station in the market on a special "demographic map." This can be done by creating a two-dimensional grid, with the vertical axis expressing the percent of males in each station's audience, and the horizontal axis expressing the median age of the audience. Once these values are known, each station can be located on the grid. Local radio market reports contain the information needed to determine these values, although a few preliminary calculations are necessary.

The most difficult calculation is determining the median age of each station's audience. The median is a descriptive statistic, much like an arithmetic average. Technically, it is the point at which half the cases in a distribution are higher and half are lower. If, for example, 50% of a station's audience is younger than 36, and 50% are older, then 36 is the median age.

To determine the median age of a station's audience, you must know the size of its audience in the various age categories reported by the ratings service. This information can be found in the "Specific Audience" section of an Arbitron report, or the "Daypart Person Estimates" of a Birch report. Table 8.6 contains that data for a single station in a single daypart. It also contains the estimated numbers of men and women, because we will want to calculate the percent of males who are listening as well. All in all, the station has 43,200 listeners in an AQH. Because radio books report audiences in 100s, it is more convenient to record that as 432 in the table. The number of listeners 65+ must be inferred from

TABLE 8.6
Calculating Media Age and Gender of Station Audience

Age Group	Male	Female	Group Frequency	Cumulative Frequency
12–17	?	?	23	23
18–24	29	50	79	102
25–34	63	41	104	206
35–44	43	60	103	309
45–54	35	27	62	371
55–64	20	17	37	408
65+	8	16	24	432
Total 12+	?	?	432	—
Total 18+	198	211	409	—
Percent M—F[a]	48%	52%		

[a] Because radio market reports do not ordinarily report the gender of persons 12 to 17, the male to female breakdown for a station's audience must be determined on the basis of those 18 and older. In this case, there are 409(00) persons 18+ in the audience.

the difference between the total audience 12+, and the sum of all other categories (i.e., 432 − 408 = 24).

The median can now be located in the following way. First, figure out the cumulative frequency. This is shown in the column on the far right-hand side of the table. Second, divide the total size of the audience in half. In this case, it is 216 (i.e., 432/2 = 216). Third, look at the cumulative distribution and find the age category in which the 216th case falls. We know that 206 people are 34.5 or younger, and that there are 103 people in the next oldest group. Therefore, the 216th case must be between 34.5 and 44.5. Fourth, locate the 216th case by interpolating within that age group. To do that, assume that the ages of the 103 people in that group are evenly distributed. To locate the median, we must move 10 cases deep into the next age group. Stated differently, we need to go 10/103 of the way into a 10-year span. That translates into .97 years (i.e., 10/103 × 10 = .97). Add that to the lower limit of the category and, bingo, the median age is 35.47 (i.e., 34.5 + .97 = 35.47).

This procedure sounds more burdensome than it really is. Once you get the hang of it, it can be done with relative ease. It is simply a way to reduce a great deal of information about the age of the station's audience into a single number. Similarly, the gender of the audience is reduced to a single number by using the male/female breakdowns in each category. (Note that gender is not reported for teens, so we use only the information on those 18+.) In the example just given, the audience was 48% male. These two numbers become coordinates that allow us to plot a point on a two-dimensional grid. Figure 8.4 shows how the station in the example above, call it WEXP, and a number of other stations might look.

Figure 8.4 is a fairly typical array of stations with different formats. As you can see, station audiences can vary widely in terms of the type of listener they attract. A "hard rock" station and a "contemporary hits" station will both tend to have young listeners, but they typically have different appeals for young men and women. A "beautiful music" station tends to attract much older listeners. In fact, some music syndicators package radio formats designed to appeal to very specific demographics. These pronounced differences in audience composition are why it is possible for different stations to be number 1 with different categories of listeners.

This kind of demographic mapping can be used by programmers in a number of ways. For example, it can help identify "holes" in the market by drawing attention to segments of the population that are unserved by a radio station. It can also offer a different way to look at the positioning of stations in a market, and how they do or do not compete for the same type of listeners. By creating maps for different dayparts, the programmer can look for shifts in audience compostion. A news and information station

ANALYSIS OF GROSS MEASURES

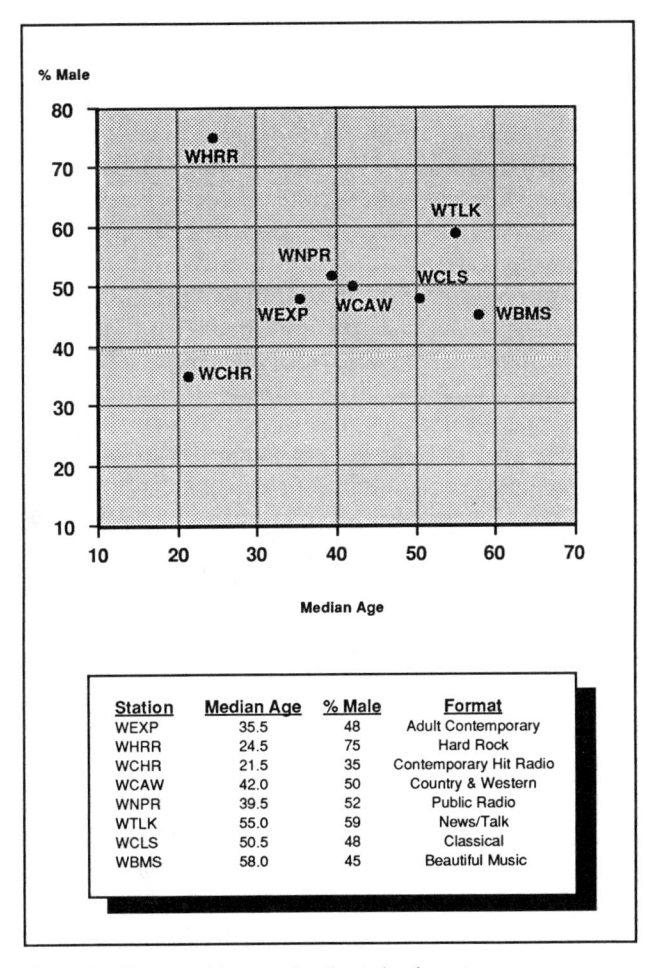

Station	Median Age	% Male	Format
WEXP	35.5	48	Adult Contemporary
WHRR	24.5	75	Hard Rock
WCHR	21.5	35	Contemporary Hit Radio
WCAW	42.0	50	Country & Western
WNPR	39.5	52	Public Radio
WTLK	55.0	59	News/Talk
WCLS	50.5	48	Classical
WBMS	58.0	45	Beautiful Music

FIG. 8.4. Demographic map of radio station formats.

might have much broader appeal in AM drive time, than in other dayparts. That could suggest changing programming strategies throughout the day.

A number of cautions in the interpretation of the map should, however, be kept in mind. First, it tells the analyst nothing about the size of the audiences involved. There may be a hole in the market because there are relatively few people of a particular type. Some markets, for example, have very old populations, others do not. Similarly, the map reports no information of the size of the station's audience, only its composition. Second, the analyst should remember that different types of listeners may be more valuable to advertisers, and hence constitute a more desirable audience. This could be attributable to their demographic composition, or the fact that many listen exclusively to their favorite station. Third, just because two or more stations occupy the same space on the map, it does not mean that they will share an audience. A country/western station and a public radio station will often fall side-by-side on a map, but typically they have very little cross-over audience. The tendency of listeners to move back and forth between stations can be more accurately studied by using the audience duplication section of the ratings book. We discuss such cumulative measurements latter. Finally, remember that age and gender are not the only factors that might be related to station preferences. The map could look quite different if ethnicity or education were among the dimensions. Just such a map can be constructed when other demographic variables such as education and income are available.

Ratings comparisons are also made longitudinally, over several points in time. In fact, most local market ratings reports will include data from previous sweeps under the heading of ratings "trends." There could be any number of reasons for looking at audience trends. A radio programmer might want to determine how a format change had altered the composition of the station's audience, perhaps producing a series of maps like Fig. 8.4. A financial analyst might want to want to examine the ratings history of a particular media property. A policymaker or economist might want to study patterns of audience diversion to assess the competitive positions of old and new media. A social scientist might examine potential changes in the social or political impact of the media.

Figure 8.5 is an example of the latter which attempts to gain insight into political communication through an analysis of ratings data. Some have argued that the president's ability to appear on the major networks and command vast television audiences is an important element of presidential power. However, it may be that the increasing availability of new media like cable and VCRs erode that power by giving the audience viewing alternatives. Foote (1988), examined that possibility by comparing the audience for presidential news conferences and addresses across time. He found that audiences shares for presidential broadcasts had

FIG. 8.5. Audience share trends for presidential broadcasts (from Foote, 1988. Reprinted with permission of the *Journal of Broadcasting and Electronic Media* © Broadcast Education Association).

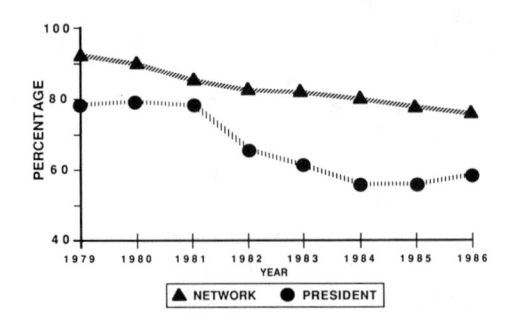

diminished over the years, far more than overall network shares. Even the "great communicator," Ronald Reagan, seemed to drive the audience away when he appeared on all three networks.

As you can surely tell by now, there are any number of ways to manipulate and compare estimates of audience size and composition. In fact, because the major ratings services now produce electronic ratings books that are read by a computer, the time involved in sorting and ranking ratings data can be drastically reduced. As we noted in chapter 6, both Arbitron and Nielsen now sell software packages that will manipulate their data in a variety of ways. Some private vendors have also produced software that can turn ratings data into bar graphs, pie charts, and so on.

Although most of these developments are a boon to ratings analysts, a number of cautions should be exercised in either producing or consuming all the comparative statistics. Do not let the computational power or colorful graphics of these programs blind you to what you are actually dealing with. Remember that all gross measures of the audience are simply estimates based on sample information. Keep the following points in mind:

- Once again, be alert to the size of the sample used to create a rating or make an audience projection. One consequence of zeroing in on a narrowly defined target audience is that the actual number of people on which estimates are based becomes quite small. That will increase the sampling error surrounding the audience estimates. A national peoplemeter sample might be big enough to breakout the audience into narrow subsets. It does not follow that local market samples, which are smaller, can provide similar target estimates of equal accuracy.
- Techniques like indexing and calculating target efficiency can have you taking percentages of percentages, which tends to obscure the original values on which they are based. For example, comparing a 1.4 rating to a .7 rating produces the same index number (i.e., 200) as comparing a 14.0 to a 7.0. Sampling error, however, will be much larger relative to the smaller ratings. This means you should have less confidence in the first index value because even slight variations in its component parts could cause it to fluctuate wildly. The fact that the second index number is more reliable is not readily apparent with this sort of data reduction.
- Keep it simple. The ability to manipulate numbers in a multitude of ways does not necessarily mean that it is a good idea. This is true for two reasons. First, the more twists and turns there are in the analysis, the more likely you are to loose sight of what you are actually doing

to the data. In other words, you are more likely to make a conceptual or computational error. Second, even if your work is flawless, more complex manipulations are harder to explain to the consumers of your research. You may understand how some special index was created, but that does not mean that a media buyer will have the time or inclination to sit through the explanation.

PREDICTION AND EXPLANATION

Most people schooled in quantitative research and theory are familiar with the concepts of prediction and explanation. In fact, the major reason for developing social scientific theory is to allow us to explain and/or predict the events we observe in the social world around us. In the context of ratings research, we can use the theories of audience behavior we developed in chapter 7 to help explain and predict gross measures of audience size and composition. Although prediction and explanation are certainly of interest to social theorists, they are not mere academic exercises. Indeed, predicting audience ratings is one of the principle activities of industry users.

It is important to remember that all ratings data are historical. They describe something that has already happened. It is equally important to remember that the buying and selling of audiences is always conducted in anticipation of future events. Although it is certainly useful to know which program had the largest audience last week, what really determines the allocation of advertising dollars is an expectation of who will have the largest audience next week or next season. Hence, ratings analysts who are involved in sales and advertising often spend a considerable portion of their time trying to predict ratings.

In the parlance of the industry, the job of predicting ratings is sometimes called *pre-buy analysis*. The buyer and seller of advertising time must each estimate the audience that will be delivered by a specific media schedule. The standard method of prediction is a two-stage process.

In the first stage, the analyst estimates the size of the total audience, as reflected in HUT or PUT levels, at the time a spot is to air. This is largely a matter of understanding audience availability. You will recall from our discussion in chapter 7 that total audience size is generally quite predictable. It varies by hour of the day, day of the week, and week of the year. It can also be affected by extremes in the weather (e.g., snow storms, heat waves, etc.), although these are obviously harder to know far in advance.

The simplest way to predict the total audience level for a future point in time is to assume it will be the same as it was exactly 1 year ago. This takes hourly, daily, and seasonal variations into account. A somewhat more involved procedure is to look at HUT/PUT over a period of months or years. By doing so, the analyst may identify long-term trends, or aberrations, in audience levels that would affect his or her judgment about jfuture HUT levels. For instance, a 4-year average of HUT levels for a given hour, day, or month, might produce a more stable estimate of future HUTs than looking only at last year, which could have been atypical. In fact, to determine audience levels during months that are measured, HUT levels should be interpolated by averaging data from sweeps before and after the month in question.

In the second stage, the analyst must project the share of audience that the station or program will achieve. Here, the simplest approach is to assume that an audience share will be the same as it was during the last measurement period. Of course, a number of factors can affect audience shares, and an analyst must take these into account. Programming changes can have a dramatic effect on audience shares. In radio, rival stations may have changed formats making them more or less appealing to some segment of the market. In television, a competing station might be counter programming more effectively than in the past. Less dramatic, long-term trends might also be at work. Perhaps cable penetration has caused a gradual erosion in audience shares that is likely to continue in the near future. Just as in estimating HUT levels, making comparisons across several measurement periods might reveal subtle shifts that would otherwise go unnoticed.

Once total audience levels and specific audience shares have been estimated, predicting an audience rating is simple. Multiply the HUT level you expect by the projected audience share, and you have a predicted rating. This formula is summarized as follows.

$$\text{Estimated HUT} \times \text{Projected Share (\%)} = \text{Predicted Rating}$$

In effect, it simply codifies the conventional wisdom expressed in Paul Klein's (1971) "theory" of the "least objectional program." That is, exposure is best thought of as a two-stage process in which an already available audience decides which station or program to watch. The procedure to predict ratings for specific demographic subsets of the audience is the same, except that you must estimate the appropriate PUT level (e.g., men 18–49) and determine the program's likely share among that audience subset. In either case, there are now a number of computer programs marketed by the ratings companies and independent vendors that perform such pre-buy analyses.

Although these formulas and computer programs are useful, remember that predicting audience ratings is not an exact science. It involves experience, intuition, and an understanding of the factors that affect audience size. Unfortunately, we can only offer help in the last category. Our advice would be to consider the model of audience behavior in chapter 7. Systematically work your way through the structural- and individual-level factors that are likely to affect audience size, and begin to test them against your own experience. Sometimes that will lead you to make modifications that "just seem to work."

For instance, one of the most difficult, and high stakes, occasions for predicting ratings occurs during upfront market in network television. Major advertising agencies and network sales executives must try to anticipate how the fall line-ups will perform. This is especially tricky because many programs are new, and have no track record to depend on. At least one major agency has found, through experience, that it can predict network ratings more accurately if it bases those predictions not on total HUT levels, but on network-only HUT levels. In other words, the first stage in the process is to estimate the total number viewers who will be watching broadcast network television. Why this results in better predictions is not entirely clear, it just seems to work.

Armed with share projections and predicted ratings for various segments of the audience, buyers and sellers negotiate a contract. Obviously, sellers are inclined to be optimistic about their ratings prospects, whereas buyers tend to be more conservative. In fact, the ratings projections that each brings to the table may be colored by the need to stake out a negotiating position. Eventually a deal is struck. Because most spot buys involve a schedule of several spots, the sum total of audience to be delivered in usually expressed in GRPs.

After a schedule of spots has run, both buyers and sellers will want to know how well they did. The process of evaluating ratings predictions is called *post-buy analysis*. In markets that are continuously measured, it is possible to know exactly how well programs performed when a spot aired. Most local markets, however, are only swept during certain months. Consequently, precise data on ratings performance may not be available. Table 8.7 identifies the sweeps that are traditionally used for post-buy analysis in different months. The point is to use the best available data for evaluative purposes.

With the schedule of spots in one hand and the actual ratings in the other, the schedule is "re-rated." For example, it may be that the original contract anticipated 200 GRPs, and than the actual audience delivered totaled 210 GRP's. If that was true, the media buyer did better that expected. Of course, the opposite could have occurred, resulting in an audience deficiency. In upfront deals, networks have traditionally made up such deficiencies by running extra spots. More often, however, it is simply the media buyer's bad luck.

Questions are often raised about how accurate ratings predictions are,

ANALYSIS OF GROSS MEASURES

TABLE 8.7
Sweeps Used for Post-Buy Analysis[a]

Sweeps Period	Months Covered for Post-Buy Analysis
February	January–February–March
May	April–May–June
July	July–August–September
November	October–November–December

[a] This schedule for post-buy analysis assumes the market is swept four times a year. Additional sweeps in January, March, and October would, if available, be used for post-buy analysis in January, March–April, and September–October, respectively.

or should be. The standard practice in the industry has been to view delivered audience levels within plus or minus 10% of the predicted levels as acceptable. There are three sources of error that can cause such discrepancies—forecasting error, strategic error, and sampling error. Forecasting errors are usually the first that come to mind. Included here are errors of judgment and prediction. For example, the analyst might not have properly gauged a trend in HUT levels or foreseen the success of some programming strategy. Strategic errors are those deliberately introduced into the process at the time of contractual negotiations. A researcher might, for instance, honestly believe that a particular program will deliver a 10 rating. The person selling the program, however, might believe that it could be sold at 12, if a projection justified that number. To make a more profitable deal, the projection is knowingly distorted.

Sampling error can also affect the accuracy of ratings predictions, and should serve to remind us once again that these numbers are simply estimates based on sample information. As we saw in chapter 5, any rating is associated with a certain amount of sampling error. The larger the sample on which the rating is based, the lower is the associated error. Further, the error surrounding small ratings tends to be rather large relative to the size of the rating itself. The same logic can be applied to a schedule of spots as expressed in GRPs. In the mid-1980s, Arbitron did an extensive study of the error (Jaffe, 1985) in GRP estimates. The principle conclusions were as follows.

- GRPs based on larger effective sample bases (ESB) had smaller standard errors. In effect this means that GRPs based on large audience segments (e.g., men 18+) are more stable than those based on smaller segments (e.g., men 18–34). It also means that larger markets, which tend to have larger samples, will generally have less error than small markets.
- The higher the pairwise correlation between programs in the schedule, the higher was the standard error. In other words, when there is

a high level of audience duplication between spots in the schedule, there is a higher probability of error. This happens because high duplication implies that the same people tend to be represented in each program rating, thereby reducing the scope of the sample on which GRPs are based.
- For a given GRP level, a schedule with relatively few highly rated spots was less prone to error that a schedule with many low-rated spots.
- All things being equal, the larger the schedule in terms of GRPs, the larger was the size of absolute standard error, but the smaller was the size of relative standard error.

As a practical matter, all this means that a post-buy analysis is more likely to find results within the plus or minus 10% criterion if GRPs are based on relatively large segments of the market, and programs or stations with relatively high ratings. A match of pre-buy predictions to post-buy ratings is less likely if GRPs are based on small ratings among small audience segments, even if forecasting and strategic error are nonexistent. As increased competition fragments radio and television audiences, and as advertisers try to target increasingly precise market segments, this problem of sampling error is likely to cause more post-buy results to fall outside the 10% range.

The method for predicting ratings we have described thus far is fairly straight-forward, requires relatively little in the way of statistical manipulations, and depends heavily on intuition and expert judgment. There have, however, been a number of efforts to model program ratings in the form of mathematical equations. With either approach, the underlying theory of audience behavior is much the same. In attempts to model the ratings, however, many expert judgments are replaced by empirically determined, quantitative relationships.

Gensch and Shaman (1980) developed a model that used a trigonometric time series to estimate accurately the number of viewers of network television at any point in time. Consistent with our earlier discussions, they discovered that total audience size was not dependent on available program content, but rather a function of time of day and seasonality. Once the size of the available audience was predicted, the second stage in the process, determining each program's share of audience, was modeled independently. This, of course is analogous to the standard method of prediction in the industry.

Rust and Alpert (1984), Horen (1980), and others have concentrated on how the available audience is distributed across program options. Here factors such as lead-in effects, counter programming, a program ratings history, program type, and the like are used to produce share estimates. In addition to these general ratings models, more specialized models have been tested. Litman (1979), for example, used data on the age and box office receipts of movies to predict their ratings on television. Multiple regression techniques are often used in such attempts to model program ratings. The dependent variable is typically program ratings or shares. Predictor variables can include lead-in ratings, program type, theatrical rentals, or whatever is theoretically justified. With information on these predictor variables entered into the equation, it is often possible to explain very substantial amounts of variation in the dependent variable. (For a more complete and technical discussion of these attempts at modeling, the reader is referred to Rust, 1986.)

Multivariate statistics can, of course, be used to do things other than predict program ratings. By emphasizing explanation rather than prediction, regression techniques can also promote a better understanding of exposure to electronic media. For example, we noted in our discussion of communications policy in chapter 3, that the FCC has had a special interest in the audience for local news programming. Social scientists, too, have questioned the ability of broadcast news to set agendas and to create an awareness of important issues. For these reasons, as well as more pragmatic concerns about the success of local programming, it seems important to understand the determinants of exposure to television news.

Traditionally, most efforts in this area have concentrated on individual-level factors such as viewer needs and preferences, as if to say, "if we only understood people's needs and motivations, we would be able to explain their viewing of TV news." Curiously, however, relatively few studies have considered the impact of structural variables on the news audience. You will recall, that among these we included audience availability, market characteristics, and programming strategies like lead-in effects.

Table 8.8 is a matrix reporting the correlations between local news

TABLE 8.8
Correlates of Local News Ratings[a]

Variables	2	3	4	5	6
1. Local news rating	.75**	.82**	.03	.25**	−.29**
2. Lead-in rating	—	.62**	−.14	.15	−.27**
3. Network news rating		—	.01	.09	−.17
4. Number of local news counter programmed			—	−.05	−.27**
5. Number of network news counter programmed				—	−.19*
6. Number of entertainment counter programmed					—

[a] Adapted from Webster and Newton (1988).
* Pearson product-moment correlation significant $p < .05$.
** Pearson product-moment correlation significant $p < .01$.

ratings around the United States and a variety of programming factors, including the rating of each station's network news. Such correlations can range in value from 0, indicating no relationship, to either plus or minus 1.0, indicating a perfect positive or negative correlation. Along the top line is the dependent variable (i.e., local news ratings) and its correlation with other independent variables.

These data indicate that there is a very strong positive correlation between local news ratings and both their lead-in ratings (i.e., 75) and their network's news ratings (i.e., .82). In other words, as lead-in and network ratings go up, so to do local news ratings, and in a fairly direct way. Further, when these two independent variables are combined with audience availability as measured in PUT levels, the three explain over 80% of the variance in local news ratings (i.e., $R^2 = .81$). Of course, as any introductory statistics book will tell you, one cannot safely infer causation from such data, but the sheer strength of these relationships is intriguing.

In our judgment, these sorts of analyses represent a fertile area for further study. In addition to the attempts at modeling program ratings cited earlier, researchers have used correlational studies of gross audience measurements to assess the success of different programming strategies (e.g., Tiedge & Ksobiech, 1986, 1987; Walker, 1988), determine the cancellation threshold of network programs (Atkin & Litman, 1986), assess the impact of media ownership on ratings performance (Parkman, 1982), and examine the role of ratings in the evolution of television program content (McDonald & Schechter, 1988). Much more can and should be done.

RELATED READINGS

Fletcher, J. E. (Ed.). (1981). *Handbook of radio and TV broadcasting: Research procedures in audience, program and revenues.* New York: Van Nostrand Reinhold.

Fletcher, J. E. (1985). *Squeezing profits out of ratings: A manual for radio managers, sales managers and programmers.* Washington, DC: National Association of Broadcasters.

Hall, R. W. (1988). *Media math: Basic techniques of media evaluation.* Lincolnwood, IL: NTC Business Books.

Sissors, J. Z., & Bumba, L. (1989). *Advertising media planning* (3rd ed.). Lincolnwood, IL: NTC Business Books.

ANALYSIS OF CUMULATIVE MEASURES   9

Cumulative measures are the second kind of audience summary that a ratings analyst has to deal with. These measures of exposure are distinguished from gross measures because they depend on tracking the audience over some period of time. Although some cumulative measures are routinely reported by the ratings services, they are less common than the gross measurements we have just reviewed. Nevertheless, we believe that thoughtful analyses of cumulative measures can provide an analyst with considerable insights into the nature of audience behavior and its possible effects.

CUMULATIVE MEASURES

We begin our discussion with a review of common cumulative measurements of the audience. A few of these appear on the printed pages of ratings books. Several more are easily, and routinely, calculated from material contained in the books. Many other cumulative measurements are possible, but require access to the appropriate database. All these are discussed here.

The most common cumulative measure of the audience is called a *cume*. A cume is the total number of different people or households who have tuned in to a station for at least 5 minutes over some longer period of time—usually a daypart, day, week, or even a month. The term *cume* is often used interchangeably with *reach* and *circulation*. When a cume is expressed as a percentage of the total possible audience, it is called a *cume rating.* When it is expressed as the actual number of people estimated to have been in the cume audience, it is called *cume persons.* These audience summaries are analogous to the ratings and projected audiences we discussed in the previous section.

Like ordinary ratings and audience projections, variations on the basic definition are common. Cumes are routinely reported for different subsets of the audience, defined both demographically and geographically. For example, Arbitron Radio reports a station's metro cume ratings for men and women of different age categories. Cume persons are also estimated within different areas of the market, like the Metro, ADI, and TSA. Regardless of how the audience subset is defined, these numbers all express the total, unduplicated, audience for a station. Each person or household in the audience can only count once in figuring the cume. It does not matter whether they listened for 5 minutes or 5 hours.

In addition to reporting cumes across various audience subsets, the ratings services will also report station cumes within different dayparts. Radio ratings books estimate a station's cume audience during so-called morning drive time (i.e., Monday through Friday, 6 a.m.–10 a.m.), afternoon drive (i.e., Monday through Friday, 3 p.m.–7 p.m.), and other standard dayparts. Cume audiences are also reported for a station's combined drive time audience (i.e., how many people listened to a station in either/or AM and PM drive time).

The period of time over which a cume audience can be determined is constrained by the measurement technique that the ratings service employs. Radio cumes cannot exceed 1 week, because the diaries used to measure radio listening are only kept for 1 week. The same is true for television cumes in diary-only markets. Barring repeated call backs, telephone recall techniques face similar limitations. Meter measurements, on the other hand, allow the ratings services to track cume audiences over longer periods of time.

In principle, household meters could produce household cumes, and peoplemeters could produce person cumes over any period of continuous operation (e.g., years). As a practical matter, cume audiences are rarely tracked for more than 1 month. Four-week cumes, however, are commonly reported with meter-based data. Since many TV programs only air once a week, this allows a ratings user to see how widely the show is viewed over several weeks. Nielsen Media Research generates 4-week cumes by measuring viewing during a specific minute in each telecast of the program. This *selected minute,* as it is called, is located at a different point in each episode. Nielsen simply notes the number of households viewing the selected minute in one or more telecasts, and divides that by the total number of households in-tab to produce the program's cume rating.

ANALYSIS OF CUMULATIVE MEASURES

Similar 4-week cumes can be calculated for TV stations in metered markets.

Two other variations on cumes are reported by radio ratings services. The first is called an *exclusive cume*. This is an estimate of the number of people who listen to one particular station, and nothing else, during a given daypart. All other things being equal, a large exclusive audience may be more saleable than one that can be reached over several stations. Arbitron and Birch Radio also report *cume duplication*. This is the opposite of an exclusive audience. For every pair of stations in a market, the rating services estimate the number of listeners who are in both stations' cume audiences. It is possible, therefore, to see which stations tend to have an audience in common.

The various cume estimates can be used in subsequent manipulations, sometimes combining them with gross measures of the audience, to produce different ways of looking at audience activity. One of the most common is a measure of *time spent listening* (TSL). The formula for computing TSL is as follows:

$$\frac{\text{AQH Persons for Daypart} \times \text{Number of Quarter Hrs in the Daypart}}{\text{Cume Persons for Daypart}} = \text{TSL}$$

Within any given daypart, for any given segment of the audience, determine the average quarter hour (AQH) audience for the station. This will be a projected audience reported in hundreds. Multiply that by the total number of quarter hours in the daypart. For AM or PM drive time, that is 80 quarter hours. For the largest daypart (Monday through Sunday 6 a.m.–midnight), it is 504 quarter hours. This product gives you a gross measure of the total number of person quarter hours people spent listening to the station. Dividing it by the number of people who actually listened to the station (i.e., the cume persons), tells you the average amount of time each person in the cume spent listening to the station. At this point, the average TSL is expressed as quarter hours per week, but it easy enough to translate this into hours per day, in order to make it more interpretable. Table 9.1 shows how this exercise could be done to compare several stations.

As you will note, the average amount of time listeners spend tuned in varies from station to station. All things being equal, a station would rather see larger TSL estimates than smaller TSL estimates. Of course, it is possible that a high TSL is based only on a few heavy users, whereas a station with low TSLs has very large audiences. For example, compare the first two stations on the list in Table 9.1. In a world of advertiser

TABLE 9.1.
Calculating TSL Estimates Across Stations[a]

Station	AQH Persons	×	504 Qtr Hrs	÷	Cume Persons	=	TSL QH per Week	=	TSL HR per Day
WAAA	500		252,000		3,500		72.0		2.57
WXXX	1,500		756,000		20,000		37.8		1.35
WBBB	6,500		3,276,000		40,000		81.9		2.93
WZZZ	1,000		504,000		12,000		42.0		1.50

[a] This sample calculation of TSL is based on estimated audiences Monday–Sunday from 6 a.m. to midnight. That daypart has 504 quarter hours.

support, gross audience size will ultimately be more important. Nonetheless, TSL comparisons can help change aggregated audience data into numbers that describe a typical listener, and so make them more comprehensible. Alhough TSLs are usually calculated for radio stations, analogous "time spent viewing" estimates could be derived by applying the same procedure to the AQH and cume estimates in the daypart summary of a television ratings report.

Another combination of cume and gross measurements is used to produce an assessment called *audience turnover*. The formula of audience turnover is:

$$\frac{\text{Cume Persons in a Daypart}}{\text{AQH Persons in a Daypart}} = \text{Turnover}$$

Estimates of audience turnover are intended to give the ratings user a sense of how rapidly different listeners cycle through the station's audience. A turnover ratio of 1 would mean that the same people were in the audience quarter hour after quarter hour. Although that kind of slavish devotion does not occur in the "real world," relatively low turnover ratios do indicate relatively high levels of station loyalty. Because listeners are constantly tuning into a station as others are tuning out, turnover can also be thought of as the number of "new" listeners a station must attract in a time period in order to replace those who are tuning out. As was the case with TSL estimates, however, the rate of audience turnover does not tell you anything definitive about audience size. A station with low cume and low AQH audiences could look just the same as a station with large audiences in a comparison of audience turnover.

A third, fairly common way to manipulate the cume estimates that appear in radio ratings books is to calculate what is called *recycling*. This manipulation of the data takes advantage of the fact the there are cumes reported for both morning and afternoon drive time, as well as the combination of those two dayparts. It is, therefore, possible to answer the ques-

ANALYSIS OF CUMULATIVE MEASURES

tion, "of the people who listened to a station in AM drive time, how many also listened in PM drive time?" This kind of information could, of course, be valuable to a programmer who has to decide whether he or she should air the same program in both time periods. Estimating the recycled audience is a two-step process.

First, you must determine how many people listened to the station during both dayparts. Suppose, for instance, that a station's cume audience in morning drive time was 5,000 persons. Let's further assume that the afternoon drive time audience was also reported to be 5,000. If it was exactly the same 5,000 people in both dayparts, then the combined cume would be 5,000 people as well (remember each person can only count once). If they were entirely different groups, the combined cume would be 10,000. That would mean no one listened in both dayparts. If the combined cume fell somewhere in between those extremes, say 8,000, then the number of people who listened in both the morning and the afternoon would be 2,000. This is determined by adding the cume for each individual daypart, and subtracting the combined cume (i.e., AM Cume + PM Cume − Combined AM & PM Cume = Persons who listen in both Dayparts).

Second, the number of persons who listen in both dayparts is divided by the cume persons for either the AM or PM daypart. The following formula below defines this simple operation.

$$\frac{\text{Cume Persons in Both Dayparts}}{\text{Cume Persons in One Daypart}} = \text{Recycling}$$

Essentially, this expresses the number of persons listening at both times as a percentage of those in either the morning or afternoon audience. Using the hypothetical numbers in the preceding paragraph we can see that 40% of the morning audience recycled to afternoon drive time (i.e., 2,000 / 5,000 = 40%).

Because nearly all radio stations get their largest audiences during the morning hours when people are first waking up, programers like to compare that figure with the number who listen at any other time of the day. It may also be useful to compare whether these same listeners also tune in during the weekend for example. In both television and radio, data detailing when the most people are listening can be used by the promotion department to schedule announcement about other programs and features on the stations. Thus, stations hope to "recycle" their listeners into other day-parts—this builds a larger AQH for the station.

Another way to express cumulative measures of the audience is in terms of reach and frequency. These concepts are widely used among advertisers and the people who plan media campaigns. The term *reach* means essentially the same thing a cume—that is, how many *different* people were reached. It is the total number of unduplicated audience members exposed to a particular media vehicle. Just as a broadcaster might want to know the weekly cume of his or her station, an advertiser will want to know the reach of an advertising campaign. Quite often that means counting exposures across different stations or networks. As is the case with cumes, reach can be expressed as the actual number of people or households exposed to a message, or it can be expressed as a percent of some universe. For instance, a media planner might talk about reaching 80% of the adult population with a particular ad campaign.

Unlike station cumes, which are usually based on 1 week's worth of data, reach estimates are generally made over a 4-week period. This makes it somewhat easier for a media buyer to make comparisons between the reach of a network schedule and monthly magazines.

Although reach expresses the total number of audience members who have seen or heard an ad at least once, it does not tell anything about the number of times any one individual has been exposed to the message. *Frequency* is the measure of audience that expresses how often a person is exposed. Usually, frequency is reported as the average number of exposures among those who were reached. So, for instance, a media planner might not only say that a campaign reached 80% of the population, but that it did so with an frequency of 2.5.

Interestingly, reach and frequency, which are both cumulative measures of the audience, bear a strict mathematical relationship to gross rating points (GRPs), which we dealt with in the previous section. That relationship is as follows.

$$\text{Reach} \times \text{Frequency} = \text{GRPs}$$

A campaign with a reach of 80% and a frequency of 2.5 would, therefore, generate 200 GRPs. Knowing the GRPs of a particular advertising schedule, however, does not give you precise information on the reach and frequency of a campaign. Nonetheless, the three terms are related, and some inferences about reach and frequency can be made on the basis of GRPs.

Figure 9.1 depicts the usual nature of the relationship. The left-hand column shows the reach of an advertising schedule. Along the bottom are frequency and GRPs. Generally speaking, ad schedules with low GRPs are associated with relatively high reach and low frequency. This can be seen in the fairly steep slope of the left-hand side of the curve. As the GRPs of a schedule increase, gains in reach occur at a reduced rate, whereas frequency of exposure begins to increase.

The diminishing contribution of GRPs to reach occurs because of differences in the amount of media people consume. People who watch a lot of TV, for example, are quickly reached with just a few commercials. The reach of a media schedule, therefore, increases rapidly in its early stages.

ANALYSIS OF CUMULATIVE MEASURES 221

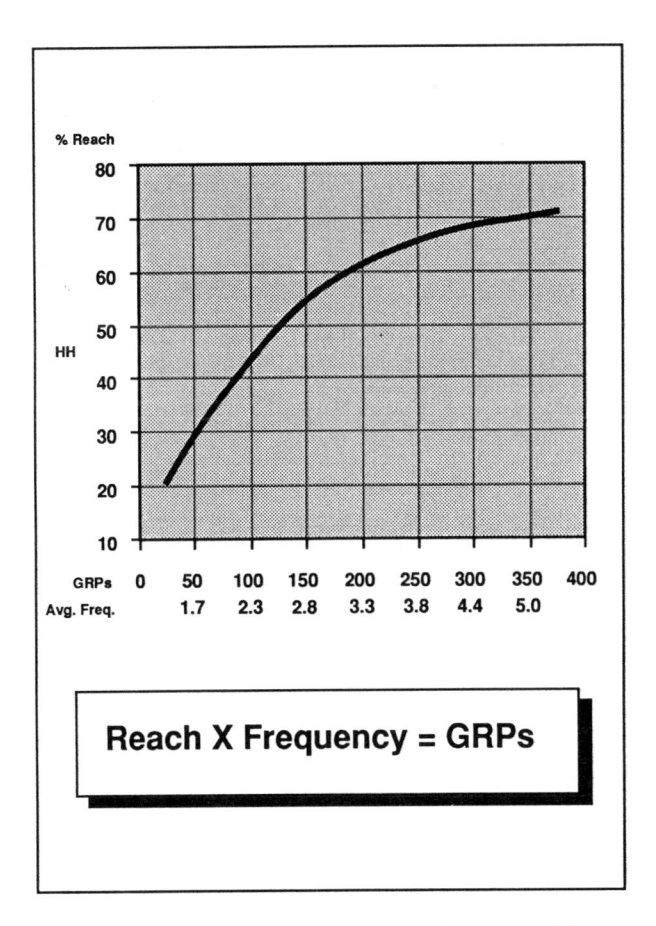

FIG. 9.1. Reach and frequency as a function of increasing GRPs.

Those who watch very little TV, however, are much harder to reach. In fact, reaching 100% of the audience is virtually impossible. Instead, as more and more GRPs are committed to an ad campaign (i.e., as more and more commercials are run), they simply increase the frequency of exposure for relatively heavy viewers. That drives up the average frequency. Across mass audiences, these patterns of reach and frequency can be predicted with a good deal of accuracy. Later in the chapter, we discuss mathematical models that are designed to do just that.

As the preceding discussion suggests, reporting the average frequency of exposure masks a lot of variation across individuals. An average frequency of 2.5 could mean that some viewers have seen an ad 15 times, whereas others have not seen it at all. It is often useful, therefore, to consider the actual distribution on which the average is based. The Niel-

TABLE B – ESTIMATES OF FREQUENCY DISTRIBUTION OF COMMERCIAL MESSAGES RECEIVED

DAYPART	NO.OF MSGS USED	AVG MSG %	4-WK REACH %	AVG NO MSGS	MM CML MSGS DLVD	1	2	3	4	5	6	7	8	9-10	11-12	13-16	17+
P-PRIME TIME	8	19.1	55.3	2.8	138	15.9	13.2	9.3	8.4	4.3	2.7	1.1	.4				
D-WEEKDAY DAYTIME	1	4.6	24.6	2.0	44	4.6	5.7	1.9	1.2	.2	.1						
A-WEEKEND DAYTIME ADULT	7	6.4	28.5	1.5	39	19.4	5.7	2.0	.5								
L-LATE FRINGE	6	5.6	22.3	1.5	30	14.0	5.7										
P-D	9	17.5	56.1	2.8	142	16.2	12.9	9.5	8.3	4.5	2.8	1.3	.5	.1			
P-A	15	13.0	61.1	3.2	177	16.7	11.7	9.5	8.0	6.3	4.0	2.3	.4	.9	.3		
P-L	14	13.3	61.0	3.0	168	16.3	13.5	9.4	8.8	5.9	3.1	2.1	1.2	.6	.1		
D-L	8	5.9	30.7	1.6	43	10.1	6.6	2.5	.9	.5							
A-L	13	5.9	42.1	1.8	69	22.7	10.8	4.5	2.6	.9	.3		.1	.1			
P-D-A	16	12.5	61.7	3.2	181	16.8	11.4	9.8	7.9	6.5	3.9	2.6	1.2	1.3	.2		
P-D-L	15	12.7	61.6	3.1	172	16.3	13.4	9.5	8.9	5.6	3.6	2.0	1.2	2.8	.2	.1	
P-A-L	21	10.9	65.7	3.5	207	16.6	12.2	9.6	7.6	7.3	4.9	2.1	1.9	2.0	.5	.1	
D-A-L	14	5.8	43.4	1.9	73	22.7	11.5	4.8	2.4	1.4	.3	.1	.1	.1			
BRAND TOTAL	22	10.6	66.1	3.5	211	16.5	12.2	9.4	8.1	7.1	4.7	3.3	1.7	2.5	.4	.2	

PHILIP MORRIS COMPANIES INC
BEER   OCT   1988   MILLER HIGH LIFE BEER
                                                                          REF CODE:134

FIG. 9.2. Brand cumulative audience (NTI) (reprinted with permission of Nielsen Television Index).

sen Television Index (NTI) reports a 4-week cumulative frequency distribution for both network programs and the brands that advertise on network television. Figure 9.2 is from NTIs *Brand Cumulative Report,* and illustrates the kind of advertising frequency one brand achieved across different dayparts. Typical of their kind, these distributions are rather lopsided, or skewed. The majority of households are exposed to far fewer advertising messages than the arithmetic average. However, a relatively small number of households, which presumably are the heavy viewers, see a great many ads.

In light of these distributions, advertisers often ask, "How many times must a commercial be seen or heard before it is effective?" Is one exposure enough for a commercial to have its intended effect, or even to be noticed? Conversely, at what point do repeated exposures become wasteful, or even counterproductive? Unfortunately, there are no simple answers to these questions. The right number of exposures depends on a variety of factors like the number of competing messages in the marketplace, and the complexity of what must be communicated. Nonetheless, many media planners will assume that an ad must be seen or heard at least three times before it can be effective. Such a minimum level of exposure is referred to as the *effective exposure* or *effective frequency*. This kind of thinking imposes a more conservative interpretation on ordinary measures of reach and frequency because those who have seen a commercial less than three times are not "effectively" reached by the campaign.

Yet another way to conceptualize cumulative audience behavior is in terms of *audience duplication*. Simply stated, analyses of audience duplication ask, "Of the people who were watching or listening at one point in time, how many were also watching or listening at another point in time?" Those points in time might be broadly defined dayparts, as is the case with recycling, or they might be brief moments, like selected minutes within different programs. In fact, audience duplication across several points in time, produces the kind of reach and frequency data just described.

Studying patterns of audience duplication is one of the most powerful and potentially illuminating techniques of analysis available to audience researchers. You may recall from chapter 7 that analyses of television audience behavior have identified such well-established patterns of duplication as inheritance effects, channel loyalty, and repeat viewing. Unfortunately, most questions of audience duplication cannot be answered by looking at the numbers published in a typical ratings report. To observe that one TV program has the same rating as its lead-in, is no assurance that the same audience watched both. Nevertheless, if one has access to the individual level data on which the ratings are based, a variety of analytical possibilities are open.

Studies of audience duplication begin with a straight-forward statistical technique called *cross-tabulation*. Cross-tabulation is described in detail in most books on research methods, and is a common procedure in statistical software packages. Cross-tab, as it is sometimes called, allows an analyst to look at the relationship between two variables. If, for instance, we had done a survey about magazine readership, we might want to identify the relationship between reader demographics and subscription (e.g., are women more or less likely to buy *Cosmopolitan* than men?). Each person's response to a question about magazine subscription could be paired with information on their gender, resulting in a cross-tabulation of those variables.

When cross-tabulation is use to study audience duplication, the analyst pairs one media-use variable with another. Suppose, for example, we had diary data on a sample of 100 people. With such data, we would be in a position to answer questions like, "Of the people who watched one situation comedy (i.e., SC1) how many also watched a second situation comedy (i.e., SC2)?" These are two behavioral variables, among a great many, contained in the data. A cross-tabulation of the two would produces a table like the one in Fig. 9.3.

The numbers along the bottom of Fig. 9.3a show that 20 people wrote "yes," they watched SC1, whereas the remaining 80 did not watch the program. The numbers in these two response categories should always add up to the total sample size. Along the far right-hand side of the table are the comparable numbers for SC2. We have again assumed that 20 people reported watching, and 80 did not. All the numbers reported along the edges, or margins, of the table are referred to as *marginals*. We should point out that when the number of people viewing a program is reported as a percentage of the total sample, that marginal is analogous to a program rating (e.g., both SC1 and SC2 have person ratings of 20).

The question, of course, is whether the same 20 people viewed both SC1 and SC2. The cross-tabulation reveals the answer in the four cells of the table. The upper left-hand cell indicates the number of people who watched SC1 and SC2. Of the 100 people in the sample, only 5 saw both programs. That is what is referred to as the *duplicated audience*. Conversely, 65 people saw neither program. When the number in any one cell is known, all numbers can be determined because the sum of each row or column must equal the appropriate marginal.

Once the size of the duplicated audience has been determined, the next problem is one of interpretation. Is what we have observed a high or low level of duplication? Could this result have happened by chance, or is there a strong relationship between the audiences for the two programs in question? To evaluate the data at hand, we need to judge our results

ANALYSIS OF CUMULATIVE MEASURES

(a)

		Viewed SC1		total
		yes	no	
Viewed SC2	yes	5	15	20
	no	15	65	80
	total	20	80	100

(b)

		Viewed SC1		total
		yes	no	
Viewed SC2	yes	O=5 E=4	O=15 E=16	20
	no	O=15 E=16	O=65 E=64	80
	total	20	80	100

FIG. 9.3. Cross-tabulation of program audiences.

against certain expectations. These expectations are either statistical or theoretical/intuitive in nature.

The statistical expectation for this sort of cross-tabulation is easy to determine. It is the level of duplication that would be observed if there was no relationship between two program audiences. In other words, because 20 people watched SC1 and 20 watched SC2, we would expect that a few people would see both, just by chance. Statisticians call this chance level of duplication the *expected frequency*. The expected frequency for any cell in the table is determined by multiplying the row marginal for that cell (R) times the column marginal (C) and dividing by the total sample (N). The formula for determining the expected frequency, then, is:

$$R \times C / N = E$$

So, for example, the expected frequency in the upper left hand cell is 4 (i.e., $20 \times 20 / 100 = 4$). Figure 9.3b shows both the observed frequency

(O) and the expected frequency (E) for the two sitcom audiences. By comparing the two, we can see that the duplicated audience we have observed is slightly larger that the laws of probability would predict (i.e., 5 > 4). Most computer programs will also run a statistical test, like *chi square*, to tell you whether the difference between observed and expected frequencies is "statistically significant."

Remember from our discussion of audience availability in chapter 7, that from one time period to another, audiences will overlap one another. For example, if 50% of the audience is watching television at one time, and 50% is watching later in the day, a certain percentage of the audience will be watching at both time. The statistical expectation of overlap is determined exactly as in the example just given, except now we are dealing with percentages. If there is no correlation between the time period audiences, then 25% will be watching at both times, just by chance (i.e., 50 × 50 / 100 = 25). You may also recall that we routinely observe a level of total audience overlap, or duplication, that exceeds chance. It is in this sort of circumstance that the second kind of expectation comes into play. An experienced analyst knows enough about audience behavior to have certain theoretical or intuitive expectations about the levels of audience duplication he or she will encounter. Consider the two sitcoms again. Suppose we knew that they were scheduled on a single channel, one after the other, at a time when other stations were broadcasting longer programs. Our understanding of inheritance effects would lead us to expect a large duplicated audience. If each show was watched by 20% of our sample, we might be surprised to find anything less than 10% of the total sample watching both. That is well above the statistical expectation of 4%. On the other hand, if the two shows were scheduled on different channels at the same time, we would expect virtually no duplication at all. In either case, we have good reason expect a strong relationship between watching SC1 and SC2.

The research and theory we reviewed in chapter 7 should give you some idea of the patterns of duplication that are known to occur in actual audience behavior. You should be alert, however, to the different ways in which information on audience duplication is reported. The number of people watching any two programs, or listening to a station at two different times, is often expressed as a percentage or a proportion. That makes it easier to compare across samples or populations of different sizes. Unfortunately, percentages can be calculated on different bases. For each cell in a simple cross-tab, each frequency could be reported as a percent of the row, the column, or the total sample.

Table 9.2 is much like the 2 × 2 table in Fig. 9.3. We have increased the size of the SC1 audience however, to make things a bit more complicated. First, you should note that changing the marginals has an impact

ANALYSIS OF CUMULATIVE MEASURES

TABLE 9.2
Cross-Tabulation of Program Audiences With Expected Frequencies and Cell Percentages

		Viewed SC1		total
		yes	no	
Viewed SC2	yes	E=6 T=6% R=30% C=20%	E=14 T=14% R=70% C=20%	20
	no	E=24 T=24% R=30% C=80%	E=56 T=56% R=70% C=80%	80
		30	70	100

on the expected frequencies (E) within each cell. When SC1 is viewed by 30, and SC2 by 20, E equals 6 (30 × 20 / 100 = 6). That change, of course, affects all the other expected frequencies. For convenience, let us also assume that we actually observe these frequencies in each box. We can express each as one of three percentages or proportions. Because our total sample size is 100, the duplicated audience is 6% of the total sample (T). Stated differently, the proportion of the audience seeing both programs in .06. We could also say that 20% (C) of the people who saw SC1 also saw SC2. Alternatively, we could say that 30% (R) of the people who saw SC2 also saw SC1.

Different expressions of audience duplication are used in different contexts. The convention is to express levels of repeatviewing as an average of row or column percentages. Because the ratings of different episodes of a program tend to be stable, these are usually quite similar. This practice results in statements like, "the average level of repeat viewing was 55%." Channel loyalty is usually indexed by studying the proportion of the total sample that sees any pair of programs broadcast on the same channel. We will have more to say about this later on when we discuss the "duplication of viewing law." Inheritance effects are studied and reported both ways. Proportions of total audience have been used to model this kind of audience flow, whereas row and column percents are often used to report typical levels of duplication between adjacent programs.

Finally, if all this were not complicated enough, we should note that ratings users sometimes make inferences about the existence of audience duplication without benefit of direct observation. For example, programs with small ratings may be aired more than once a day. Under such circumstances, it is not unusual for the rating from each airing to be totaled and sold as if it were a single program rating. This is done because; (a) many time buyers do not like to deal a lot of tiny ratings, and (b) it is assumed that no one watches the same program twice in a day, therefore, no audience duplication occurs across programs. It seems likely that as cable networks, which often repeat programming, fragment the audience, this practice may increase. Whether it is based on a sound understanding of audience behavior could and should be examined.

Similarly, some rating analysts have inferred levels of audience duplication by looking at correlations between program ratings or shares. Under this approach, it is assumed that pairs of programs with highly correlated ratings have relatively high levels of duplication, although program pairs with low correlations have low levels of duplication. For instance, researchers have examined the correlations among adjacent program audience shares (e.g., Tiedge & Ksobiech, 1986; Walker, 1988), arguing that conditions that produce high correlations indicate relatively pronounced inheritance effects. Because no direct observation of audience duplication is made, however, such correlational data is only "circumstantial evidence" of audience flow. Although this approach is clearly less desirable than studying actual levels of duplication it can, nonetheless, produce useful insights into audience behavior.

COMPARISONS

As is the case with gross measures of the audience, it is common practice to make comparisons among cumulative measures. Comparisons, after all, can provide a useful context within which to interpret the numbers. However, with cumulative measures, part of the impetus for comparing every conceivable audience subset, indexed in every imaginable way, is absent. As a practical matter, gross measures are used more extensively in buying and selling audiences than are cumulative measures. Hence, there is less pressure to demonstrate some comparative advantage, no matter how obscure. Although some cume estimates, like reach, frequency, and exclusive cumes, can certainly be useful in time sales, much of the comparative work with cumulative measures is done to realize some deeper understanding of audience behavior.

Programmers can often benefit from such insights. A radio station, for example, might wish to cultivate a small but loyal audience. Perhaps the strategy is to offer a unique, or fairly narrow, format that is intended to have strong appeal to a limited number of people in the market. A Spanish language station might have such an objective. If so, the programmer would want to consider not only gross measures of audience size, but cumulative measures as well. Is the average TSL for the station any greater than for other formats in the market—what about audience turnover and exclusive cumes? Table 9.3 shows how such comparisons might look by giving the average TSL, turnover ratio, and exclusive cume for stations with different formats.

For programming analysis it is especially useful to compare the TSL and exclusive cume of stations with the same format, similar formats, and stations that "share" their audiences. In the same way, it is helpful to analyze the programs and personalities on just one station by comparing various dayparts. The TSL can also be computed to compare men and women in various age categories to show who are the heaviest and lightest radio users in a specific market.

Certainly if a radio station was changing its format, it would want trend data on both gross and cumulative measures. For example, if an attempt was being made to broaden the station's format, or to position it closer to other competitors in the market, one might expect accompanying shifts in its cume audiences. Let's say a station changed from having the only AOR format in the market, to being one of two "classic rock" stations in the market, a programmer would be well advised to monitor the exclusive and duplicated cumes of each station. Although a change in format might reduce the station's exclusive cume, if its attempt at repositioning has been successful, it should enjoy increased duplication of audience with the other rock station. Conversely, the existing rock station might find its own exclusive cume reduced. These sorts of insights can be gleaned by comparing the performance of each station over time.

TABLE 9.3.
Time Spent Listening, Turnover Ratio, and Exclusive Cume for Various Radio Formats[a]

	TSL Hours:Mins	Turnover Ratio	Exclusive Cume
Contemporary hit	2:15	14	11%
Urban contemp/black	2:42	12	13%
AOR/classic rock	2:05	15	9%
Adult contemp/soft	1:59	16	7%
Country	2:48	11	19%
Middle of road/variety	2:21	14	12%
Jazz/new age	1:54	17	5%
Classical	1:47	18	5%
Beautiful/EZ	2:47	11	16%
Big band/nostalgia	2:42	12	14%
News/talk	2:02	16	8%
Religious	1:56	16	11%

[a]Based on Arbitron data, Spring (1989), for a total of more than 800 stations as computed in Duncan (1989).

Interesting analyses have also been performed by looking at the reach and time spent viewing of television stations. Barwise and Ehrenberg (1988) have argued that, unlike radio, television stations rarely have small, but loyal, audiences. Instead, it is almost always the case that a station that reaches a small segment of the audience is viewed by that audience only sparingly. This is sometimes labeled a *double jeopardy* effect, because a station suffers not only from having relatively few viewers, but also from having relatively disloyal or irregular viewers as well. To demonstrate the double jeopardy effect, they constructed a graph based on television ratings data in both the United States and the United Kingdom (see Fig. 9.4).

Along the horizontal axis is the weekly reach achieved by various types

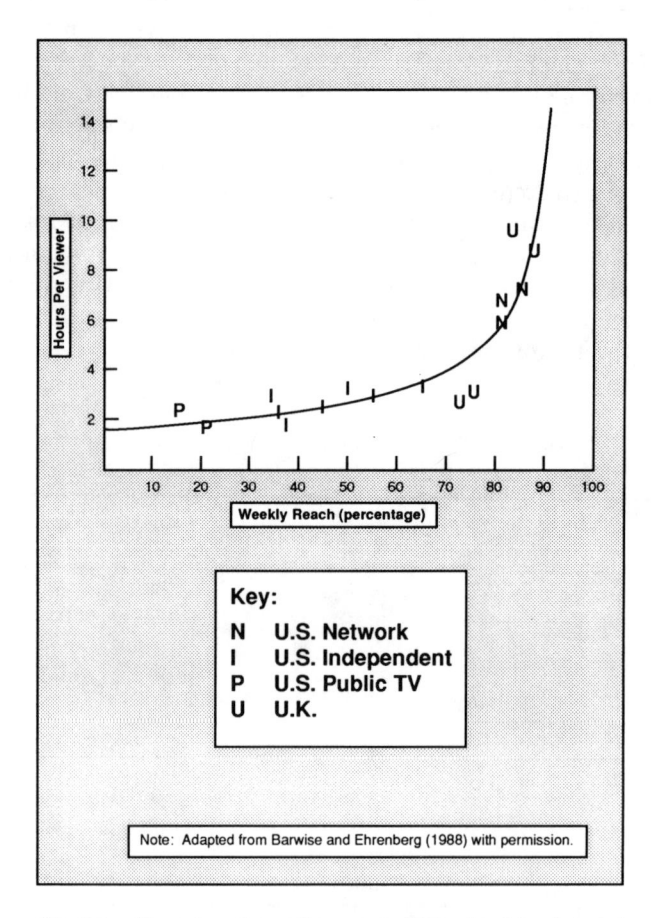

FIG. 9.4. Channel reach and time spent viewing television (adapted from Barwise & Ehrenberg, 1988, with permission).

of stations, expressed as a percent of the total audience. Along the vertical axis is the average number of hours each station's audience spent viewing the station in a week. This can be determined in the same way that TSL estimates are made in radio. As you can see, the slope of the curve is rather flat to begin with, but rises sharply as the reach of the station increases. This means that, as a rule, low levels of station reach are associated with small amounts of time viewing, but as reach moves beyond 50% or so, increased reach is associated with dramatic increases in weekly time spent viewing (TSV). Further, Barwise and Ehrenberg reported that the curve depicted in the graph is accurately summarized by the equation:

$$1.8 + \frac{\text{weekly reach \%}}{100 - \text{weekly reach \%}} = \text{TSV}$$

Aside from the fact that this pattern of station reach is easy to model with an equation, the curve is interesting for what it reveals about station audiences. As is widely known, public television stations are viewed by relatively few people (i.e., they have low weekly cumes). However, contrary to what many assume, these are not PBS loyalists who spend a lot of time with the station. Instead, they watch very little public television in a week's time. The only exceptions that Barwise and Ehrenberg report finding are for religious and minority language TV stations. These, apparently, do have audiences that are both small and loyal. Whether the double jeopardy pattern typical of most broadcast television applies to specialized channels in a multichannel environment, like cable, is yet to be determined.

As we noted in our discussion of audience duplication, many analyses are possible only with access to the complete ratings database. Barwise and Ehrenberg (1988), have published analyses of television audience behavior that illustrate some of the possibilities. For example, a research question of considerable importance to both programmers and theorists is whether people who watch one program of a type will tend to watch others of the same type. As a practical matter, the answer has implications for how to manage audience flow. Beyond its applied value, however, the answer could also support or undermine different theories of program choice that predict a demonstrable consistency of viewer preference.

Table 9.4 offers a simple way to test whether viewers demonstrate program type loyalty in their choice of programming. All *programs* have been categorized into one of seven types. These are listed across the top of the table. Down the left hand side of the table, the authors have segmented *viewers* according to which programs they have watched. Having done so, they can report how the average audience for each kind of program distributes the remainder of the TV viewing time. So, for example, the

TABLE 9.4.
Viewing of Program Types by Program Type Audiences[a]

	Percentage of their other viewing time spent watching each of seven program types below						
	1	2	3	4	5	6	7
Viewers of an average program of:							
1. Light entertainment	15	26	11	9	3	17	9
2. Light drama	15	27	12	9	3	18	9
3. Films	14	26	13	9	3	17	9
4. Sports	15	26	11	10	3	17	9
5. Drama & arts	15	26	12	9	3	18	9
6. Information	15	26	11	9	3	17	9
7. News	15	26	11	9	3	17	10
Average	15	26	12	9	3	17	9

[a]Adapted from Barwise and Ehrenberg (1988) with permission.

audience for an average news show spends 15% of its time watching light entertainment, 26% watching light drama, and so forth.

These data would seem to suggest that there are no special patterns of program type loyalty. People who watch one news program are unlikely to be "news junkies," but rather watch about as much news as everyone else. In fact, each audience segment watches about the same proportion of each program type as the other audience segments. That is to say that many viewers seem to which a fairly wide variety of different program types rather than limited their viewing to only one or two kinds of programs.

These results have been based on analyses of audience duplication in the United Kingdom. They have yet to be fully replicated in the United States, which, with many more channels, may produce a different result. What is important to note here, is that data on program audience duplication can be combined and compared so as to produce a large and intriguing picture of exposure to program content.

PREDICTION AND EXPLANATION

The audience behavior revealed in cumulative measurements can be quite predictable—at least in the statistical sense. Because we are dealing with mass behavior occurring in a relatively stable environment over a period of days or weeks, that behavior can be approximated with mathematical models, sometimes with great accuracy. This is certainly a boon to media

ANALYSIS OF CUMULATIVE MEASURES

planners attempting to orchestrate effective campaigns, especially because actual data on audience duplication is always after the fact and often hard to come by. As a result, much attention has been paid to developing techniques for predicting reach, duplication, and frequency of exposure.

The simplest model for estimating the reach of a media vehicle is given by the following equation;

$$\text{Reach} = 1 - (1 - r)^n$$

where r is the rating of the media vehicle, and n is the number of ads, or insertions, that are run in the campaign. When applying this equation, it is necessary to express the rating as a proportion (e.g., a rating of 20 = .20). Although straight-forward, this model of reach is rather limited. In the early 1960s, more sophisticated models were developed (e.g., Agostini, 1962; Metheringham, 1964) based either on binomial or beta binomial distributions. These and other models of reach are described in detail by Rust (1986).

Most advertising agencies of any size will have computer programs available to predict the reach of a media schedule based on input like GRPs. It is also common to predict reach for different demographic segments of the audience, within different dayparts. Although such programs are quite useful, the analyst should remember that these are projections, not "the truth." Baron (1988), for example, has pointed out that observed levels of reach are subject to more variation than the smooth curves of a mathematical model would suggest. Therefore, deciding among media plans on the basis of small differences in computer projections is foolish.

Although models of reach embody some assumptions about audience duplication, to predict duplication between specific pairs of programs, it is best to employ models designed for that purpose. One of the most widely used models, called the "duplication of viewing law," was developed by Goodhardt, Ehrenberg, and Collins (1987). It is expressed in the following equation:

$$r_{st} = kr_s r_t$$

where r_{st} is the proportion of the audience that sees both Programs s and t, r_s is the proportion seeing program s, r_t is the proportion seeing Program t (i.e., their ratings expressed a proportions), and k is a constant whose value must be empirically determined. When the ratings are expressed as percentages, the equation changes slightly to:

$$r_{st} = kr_s r_t / 100$$

The logic behind the duplication of viewing law is not as complicated as it might appear. In fact, it is almost exactly the same as determining an expected frequency in cross-tabulation. If we were trying to predict the percent of the entire population that saw any two programs we can begin by estimating the expected frequency. Remember that is determined by $E = R \times C / N$. If its program ratings we are dealing with, that is the same as multiplying the rating of one program (s) times the rating of another (t), and dividing by 100 (the total N as a percentage). In other words, the general equation for expected frequency becomes $r_{st} = r_s r_t /100$, when it is specifically applied to predicting audience duplication. That is exactly the same as the duplication of viewing equation, with the of exception the k coefficient.

Goodhardt and his colleagues compared the expected level of duplication with the actual, or observed, level of duplication across hundreds of program pairing. They discovered that under certain well defined circumstances, actual levels of duplication were either greater or less than chance, by a predictable amount. For example, for any pair of programs broadcast on ABC, on different days, it was the case that audience duplication exceeded chance by about 60%. In other words, people who watched one ABC program were 60% more likely than the general population to show up in the audience for another ABC program on a different day. To adapt the equation so that it accurately predicted duplication, it was necessary to introduce a new term, the *k coefficient*. If duplication exceeded chance by 60%, then the value of k would have to be 1.6.

The values of k were determined for channels in both the United States and the United Kingdom. American networks have a duplication constant of approximately 1.5 to 1.6, whereas English channels have a constant on the order of 1.7 to 1.9. These constants serve as an index of *channel loyalty*, the higher the value of k, the greater the tendency toward duplication or loyalty.

Noting deviations from levels of duplication predicted by the duplication of viewing law also serves as a way to identify unusual features in audience behavior. In effect, the law gives us an empirically grounded theoretical expectation against which to judge specific observations. One important deviation for the law is inheritance effects. That is, when the pair of programs in question is scheduled back-to-back on the same channel, the level of duplication routinely exceeds that predicted by ordinary channel loyalty. Henriksen (1985) has suggested that a more flexible model of duplication, predicting both channel loyalty and inheritance effects, could be derived from models in human ecology. His model is in the form of a linear equation.

$$\log r_{st} = \log k + (b)\log r_s + (c)\log r_t$$

The duplication of viewing law has also been criticized for treating the k coefficient as a constant. In fact, there is evidence that k varies

ANALYSIS OF CUMULATIVE MEASURES

considerably across individual program pairings (e.g., Chandon, 1976; Headen, Klompmaker, & Rust, 1979; Henriksen, 1985). The duplication of viewing law is also incapable of explicitly incorporating other factors that may effect the level duplication between program pairs.

To address these limitations, Headen, Klompmaker, and Rust (1979) proposed using a more conventional regression equation to model audience duplication across program pairs. The equation takes the general form:

$$r_{st} = b_0(b_1^{x_1})(b_2^{x_2})(b_3^{x_3})(b_4^{x_4})(r_s r_t)(e^u)$$

where r_{st} is the proportion of the population seeing Programs s and t, $r_s r_t$ is the product of the programs' ratings expressed as proportions, and x_1 through x_4 are dummy variables indicating whether the programs in a pair were on the same channel, of the same type, and so forth. As the duplication of viewing law would suggest, $r_s r_t$ is the single best predictor of audience duplication, although other factors, including similarities in program type, add significantly to explained variation in r_{st}. It might also be noted, that when using linear regressions on data such as these, it is typically necessary to perform logarithmic transformations of the audience proportions to avoid violating the assumptions that underlie the linear model.

Webster (1985) used a similar method of modeling to explain audience duplication between adjacent program pairs. However, he allowed each program rating (i.e., r_s and r_t) to enter the equation independently in order to assess the relative strength of lead-in versus lead-out effects. By doing so, he established that the ratings of the earlier program in an adjacent pair explained considerably more variation than the second program rating. Overall, a model with four predictor variables explained 85% of the variation in inheritance effects.

Frequency of exposure can also been modeled. Obviously, if it is possible to predict reach on the basis of GRPs, average frequency at a certain GRP level can be determined, because the three are directly related. However, as we noted in our discussion of those concepts, it is often useful to know the entire distribution on which an average frequency is based. In that way, judgments about effective exposure can be made. Consequently, models predicting an entire distribution of exposures have been developed. There are, in fact, a great many such models. Some are based on binomial distributions, some on multivariate extensions of beta binomial distributions. Some require information on pairwise duplication as input, some do not. (For a full discussion of these alternatives, the reader is referred to Rust, 1986.)

Goodhardt et al. (1987), have employed one such model, based on a beta binomial distribution (BBD), to predict what percent of the population will see a certain number of episodes in a series. Table 9.5 compares the

TABLE 9.5
Observed Versus Theoretical Frequency Distribution for Episodes of "Brideshead Revisited"[a]

	Number of episodes seen											
	0	1	2	3	4	5	6	7	8	9	10	11
Observed freq.	40%	17%	11%	8%	4%	3%	3%	4%	2%	4%	2%	2%
Theoretical freq.	43	14	9	7	5	5	4	4	3	2	2	2

[a]Adapted from Goodhardt, Ehrenberg, and Collins (1987) with permission.

predictions of the BBD model with actual observations for 11 episodes of the series "Brideshead Revisited." The table indicates that 40% of the population did not watch any broadcast of the series. Seventeen percent saw just one episode in 11, 11% saw two episodes, and so forth. These data are exactly like those reported by Nielsen in its *Program Cumulative Audience* reports, except they extend beyond the usual 4-week time frame. The line of numbers just below the observed frequency distribution is the prediction of the BBD model. Although there are some discrepancies, the model provides a good fit to actual patterns of audience behavior.

It should be apparent by now that cumes, reach, frequency, and audience duplication are just different ways of expressing the same underlying audience behavior. In fact, information on pairwise duplication can be used to predict frequency distributions, and frequency distributions can be translated into analogous statements about audience duplication (e.g., Barwise, 1986).

It should also be apparent that ratings data, when properly analyzed, have the potential to answer an enormous number of questions. Certainly, these include the pragmatic concerns that prompted the creation of ratings in the first place. But, as we noted in chapter 3, the data are flexible enough to address problems in public policy, economics, cultural studies, and media effects. The successful application of ratings analysis to these problems, of course, requires access to the appropriate data, and an understanding of their limitations. Perhaps most importantly, however, it requires an appreciation of the audience behavior expressed in the ratings and the factors that shape it. Only then, can we exploit the data for all the insights they might offer.

RELATED READINGS

Barwise, P., & Ehrenberg, A. (1988). *Television and its audience*. London: Sage.
Fletcher, J. E. (1985). *Squeezing profits out of ratings: A manual for radio managers, sales managers and programmers*. Washington, DC: National Association of Broadcasters.

Goodhardt, G. J., Ehrenberg, A. S. C., & Collins, M. A. (1987). *The television audience: Patterns of viewing*. Aldershot, UK: Gower.

Rust, R. T. (1986). *Advertising media models: A practical guide*. Lexington: Lexington Books.

Sissors, J. Z., & Bumba, L. (1989). *Advertising media planning* (3rd ed.). Lincolnwood, IL: NTC Business Books.

APPENDIX A:
NATIONAL RATINGS RESEARCH COMPANIES: ADDRESSES, SERVICES, AND RELATED INSTITUTIONS

A.C. Nielsen (A Dun & Bradstreet Company)
Chicago
    Nielsen Media Research
    Nielsen Plaza
    Northbrook, IL 60062
    (708) 498-6300
New York
    Nielsen Media Research
    1290 Avenue of the Americas
    New York, NY 10104
    (212) 708-7500

Nielsen Television Index

The NTI is accredited by the Electronic Media Rating Council. This national ratings service for television and cable networks provides a variety of reports, the most common of which are described here.

Nielsen National TV Ratings, commonly referred to as the "Pocketpiece" is issued weekly, 52 weeks a year. It contains household and person audience estimates for sponsored network television programs. See chapter 6 for a typical page from this report.

NTI Planner's Report is issued annually contains estimates of TV usage and network program audiences useful in planning network television buys.

Up-Front Buying Guide (UFBG) provides quarterly estimates of household and person TV usage useful in planning network television buys.

APPENDIX A: RATINGS RESEARCH COMPANIES 239

Households Using Television Summary (HUT) issued quarterly, provides estimates of U.S. television household usage for each half-hour time period week and day of the week. It is useful in predicting future audiences for network or spot television buys.

Market Section Audiences (MSA) issued 12 times a year, it provides television audience estimates by a number of less commonly reported market categories including education of head of household, cable subscription, household income, time zone, and selected upper demographics.

Program Cumulative Audience (PCA) published five times annually, estimates the percent of households that have seen a program one, two, three, four, or more times over a 4-week period. It is useful in assessing repeat viewing and frequency of exposure.

Brand Cumulative Audiences Report (BCA) provides estimates of cumulative audiences (reach and frequency) to the advertising schedules of NTI client brands.

National Audience Demographics Report (NAD) is a two-volume report issued 12 times a year, breaking down audiences in a wide variety of age, gender, and household characteristic combinations. *Household & Persons Cost Per Thousand Report (CPT)* is issued 12 times a year. It provides various CPMs for 30-second commercials useful in judging the efficiency of a media plan.

Tracking Report (Households and Persons) is updated monthly and it tracks the performance of program over a 2-year period.

Dailies Plus is a computer-delivered service that provides household and person ratings the day after a network prime-time program is aired, and syndicated program performance data on a weekly basis. Dailies plus data are downloaded to a client's computer.

Person's Cume Facility is an on-line computer service that allows clients to analyze reach, frequency, and audience duplication for customized schedules.

Nielsen Station Index

The NSI is accredited by the Electronic Media Rating Council. Its major services are summarized here.

Viewers in Profile is the name of the local market reports NSI issues. In most markets these are based on diary data and are issued four times a year. See chapter 6 for a discussion of these and metered services available in larger markets.

Report on Syndicated Programs (ROSP) combines local market information to assess the performance of syndicated programs across markets.

Cassandra is a computer software system capable of producing more customized analyses of syndicated programs across local markets.

Network Programs by DMA is issued quarterly and combines local market data to assess network program performance across markets.

County/Coverage Study breaks out audience estimates on a county by county basis and is useful in pinpointing station strengths and media planning.

NSI Plus is an on-line computer facility that allows clients to assess audiences in customized market areas (e.g., by zip code), and conduct cumulative analyses including reach, frequency, and audience flow.

In addition, NSI had a variety of software designed to run on personal computers and facilitate the analysis and presentation of its data. These programs include *Audience Analyst, Spotbuyer, Postbuy Reporter, Metered Market Data System,* and *TV Conquest.*

Nielsen Homevideo Index

The Homevideo Index specializes in estimating audience use of newer forms of televised media like cable and VCRs. Its major reports are discussed here.

Cable National Audience Demographics Report (CNAD), based on Nielsen peoplemeters, this provides estimates of the national audience for major cable networks across various demographic groupings and facilitates direct comparisons of those networks.

Cable Activity Report (NCAR) is issued quarterly and it provides audience ratings, shares, and cumes for both cable and broadcast television networks.

In addition, Nielsen has offered many specialized reports for cable networks and superstations, each based on a given service's universe of potential viewers.

Nielsen Syndication Service

In response to the growing syndication market Nielsen created a special service. Its principle report is the *NSS Pocketpiece.* Issued weekly for all 52 weeks a year, it is modeled after the network pocketpiece and provides household and persons estimates for programs distributed by NSS clients.

Arbitron Ratings Company (A Control Data Company)
   New York
      142 West 57th Street
      New York, NY 10019
      (212) 887-1300
   Washington
      The Arbitron Building
      Laurel, MD 20707
      (301) 497-4742

Television Services

Arbitron provides both television and radio audience estimates. The principle television reports and services is offers are described here.

APPENDIX A: RATINGS RESEARCH COMPANIES

Television Market Reports are accredited by the Electronic Media Rating Council and provide audience estimates for local markets areas. Most market reports are based on diary data and published four times a year. See chapter 6 for a description of these and the metered services avialable in larger markets.

ScanAmerica is a syndicated research service currently available in Denver. It combines peoplemeter data with product purchase information collected in home with a pen that reads universal product codes.

County Coverage Study is published yearly and provides audience station share and cume estimates within counties. Cable/noncable breakdowns are also available. These are useful in media planning and determining stations that are significantly viewed.

Arbitron offers what it refers to as a "Broadcast Management Series" that includes reports such as the *ADI Market Guide, Network Program Analysis,* and *Syndicated Program Analysis.*

Arbitron Information on Demand (AID) is an on-line computer system that allows clients to produce audience estimates for customized market areas or demographics, and analyze reach, frequency, and audience flow.

Arbitron also offers a number of programs designed to run on personal computers and aid in the analysis and presentation of its data. These programs include *TV Maximiser, Custom Tarket AID (CTA), Product Target AID (PTA),* and *Arbitrends II.*

Radio Services

Arbitron also offers a range of services dealing with radio listenership. The major services in this category follow here.

Radio Market Reports are published for over 250 market areas at least once a year, in the spring. Larger markets are reported on more often. Arbitron's radio market reports are accredited by the Electronic Media Rating Council. See chapter 6 for a description of these reports.

County Coverage reports, updated annually, provides an estimate of station listenership in every county in the United States.

Nationwide is a report that provides ADI audience estimates for wired and unwired radio networks. It is based on local market data collected in Spring, and published once a year.

Mechanical Diary provides clients with a print out of all the listening entries made by diary keepers who mention the client's station.

Arbitron Information on Demand (AID) like the television service is an on-line computer facility that provides access to the diary database and allows customized analyses of listenership.

Arbitron also has PC based programs to manipulate and analyze its data and data from other vendors, these services include *Arbitrends, Radio FasTraQ, CrossTraQ,* and *Programmer's Package.*

Birch/Scarborough Research
Coral Springs
12350 NW 39th Street
Coral Springs, FL 33065
(305) 753-6043
New York
560 Sylvan Avenue
Englewood Cliffs, NJ 07632
(201) 871-0011

Birch Radio Quarterly Summary Reports (QSR), available in over 100 markets, are produced one a quarter and include the most recent 3-month sample data. They include station AQH and cume estimates for a various demographic groups.

Birch Radio Monthly Trend Reports (MTR) are produced monthly for over 100 markets based on a rolling 2-month average.

Birch Radio Standard Market Reports (SMR) are produced annally or semi-annually in small to medium-sized markets. They contain much of the same information as QSRs.

Condensed Market Reports (CMR) provide limited demographic data and are used only in small markets.

Birch also offers PC based services for its clients, these include program called *BirchPlus* and *Radio Spotbuyer*. On-line access to Birch data is also available through third party vendors.

Electronic Media Rating Council, Inc.
420 Lexington Avenue
New York, NY 10017
(212) 687-7733

Supported by a coalition of industry groups, EMRC audits research procedures and provides an accreditation to those services that meet its standards. Not all ratings reports have, or have sought, EMRC accreditation. EMRC also provides a mechanism for arbitrating disputes between ratings services and clients.

Statistical Research, Inc.
111 Prospect Street
Westfield, NJ 07090
(201) 654-4000

Statistical Research, Inc. (SRI) offers a ratings service called Radio's All Dimension Audience Research, more commonly known as *RADAR*. RADAR, which is accredited by the EMRC, provides various national measurements of radio listenership. RADAR reports are published twice a year in three volumes. Volume 1 deals with overall radio usage and estimates the audience for both AM and FM service. Volumes 2 and 3 provide estimates of radio network audiences to commericals.

RADAR On-Line (ROL) is a computerized service that allows SRI clients direct access to the RADAR database via telephone lines.

APPENDIX B: GLOSSARY

AAAA (American Association of Advertising Agencies): a trade association of U.S. advertising agencies.

Active Audience: a term given to viewers who are highly selective about the programming they choose to watch. Active audiences are sometimes defined as those who turn a set on only to watch favored programs, and turn the set off when those programs are unavailable. See LOP, and passive audience.

Adjacency: an opening for an ad that is immediately before or after a specific program.

ADI (Area of Dominant Influence): the term used by Arbitron to describe a specific market area. Every county in the United States is assigned to one, and only one, ADI. See DMA.

Adjacent ADI: market areas immediately next to any particular ADI. Arbitron will report significant viewing in neighboring ADI's in its local market reports.

Advance Rating: audience estimates provided by a ratings firm prior to their actual publication.

Advertising Agency: a company that prepares and places advertising for its clients. Agencies typically have media departments that specialize in planning, buying, and evaluating advertising time.

Affiliate: a broadcast station that has a contractual agreement to air network programming.

AID (Arbitron Information on Demand): an on-line computerized service offered by Arbitron that permits customized ratings analysis including reach, frequency, and audience flow.

AGB (Audits of Great Britain): a marketing research firm that provides ratings in several countries around the world, including estimates based on peoplemeter data.

AMOL (Automated Measurement of Lineups): a system that electronically

APPENDIX B: GLOSSARY

determines the broadcast network programs actually aired in a local market. See Preempt.

ANA (Association of National Advertisers): a trade organization of major national advertisers responsible for creating the first broadcast ratings service. See CAB.

AQH (Average Quarter Hour): the standard unit of time for reporting average audience estimates (e.g., AQH rating, AQH share) within specified dayparts.

ARB (Audience Research Bureau): a ratings company established in 1949 that was the predecessor of the Arbitron Company.

Arbitrends: a monthly series of reports offered by Arbitron that provides a rolling average based on 3 months data.

Arbitron: a major supplier of local market ratings in both radio and television.

Area Probability Sample: a kind of random sample in which geographic areas are considered for selection in some stage of the sampling process. See probability sample, cluster sample.

ARF (Advertising Research Foundation): trade organization of advertising and marketing research professionals advancing the practice and validity of advertising research.

Ascription: a procedure for resolving confused or inaccurate diary entries such as reports of listening to nonexistent stations. See flip.

Audience Flow: the extent to which audiences persist from one program or time period to the next. See audience duplication, inheritance effects.

Audience Duplication: a cumulative measure of the audience that describes the extent to which audience members for one program, or station are also in the audience of another program or station. See audience flow, channel loyalty, inheritance effect, repeat viewing, recycling.

Audience Deficiency (AD): a failure to deliver the numbers and kinds of audiences agreed to in a contract between time sellers and buyers. Sellers will often remedy ADs by running extra commercials.

Audience Fragmentation: a phenomenon in which the total audience for a medium is widely distributed across a large number of program services. Cable is said to fragment the television audience, resulting in a decreased average audience share for each channel.

Audience Polarization: a phenomenon associated with audience fragmentation, in which the audiences for channels or stations use them more intensively than an average audience member. See channel loyalty, channel repertoire.

Audience Projection: See projected audience.

Audience Turnover: a phenomenon of audience behavior usually expressed as the ratio of a station's cumulative audience to its average quarter hour audience.

Audimeter: Nielsen's name for several generations of the metering device it has used to record set tuning. See SIA.

Available Audience: the total number of people who are, realistically, in a position to use a medium at any point in time. It is often operationally defined as those actually using the medium (i.e., PUT or PUR levels)

Availabilities: advertising time slots that are unsold, and therefore available for sale. Sometimes called *avails*.

Average: a measure of central tendency that expresses what is typical about a particular variable. An arithmetic average is usually called a *mean*. See mean, median.

Average Audience Rating: the rating of a station or program at an average point in time within some specified period of time. Metered data, for example, allow reports of audience size an average minute during a television program.

Away-from-Home Listening: estimates of radio listening that occur outside the home. Such listening usually takes place in a car or place of work.

BAR (Broadcast Advertiser Reports): a syndicated research service that monitors commercial activity, owned by Arbitron.

Barter: a type of program syndication in which the cost of the programming is reduced, sometimes to zero, because it contains advertising.

Basic Cable: the programming services provided by a cable system for the lowest of its monthly charges. These services typically include local television signals, advertiser-supported cable networks, and local access.

Birch: a research company providing syndicated radio rating reports in most U.S. markets.

Block Programming: the practice of scheduling similar programs in sequence in order to promote audience flow. See inheritance effect.

Bounce: the tendency of a station's ratings to fluctuate from one market report to the next due to sampling error, rather than real changes in audience behavior. Bounce is most noticeable for stations with low ratings.

BPME (Broadcast Promotion & Marketing Executives): a trade organization for media professionals specializing in promotions and marketing.

Broadband: a term describing the channel capacity of a distribution system. A common label for multichannel cable service, it is also applied to digital networks capable of delivering full motion video. See cable system.

Buffer Sample: a supplemental sample used by a rating company in the event that the originally designated sample is insufficient due to unexpectedly low cooperation rates.

CAB (Cabletelevision Advertising Bureau): a trade organization formed to promote advertising on cable television.

CAB (Cooperative Analysis of Broadcasting): the first ratings company. Formed in 1930 by Archibald Crossley, it ended operations in 1946.

Cable System: a video distribution system that uses coaxial cable and optical fiber to deliver multichannel service to households within a geographically defined franchise area.

Cable Penetration: the extent to which households in a given market subscribe to cable service. Typically expressed as the percent of TV households subscribing to basic cable.

Call-Back: the practice in attempting to interview someone in a survey sample who was not contacted or interviewed on an earlier try. The number of call-back attempts is an important determinant of response rates and nonresponse error. See nonresponse error.

Calibration: the controversial practice of applying mathematical corrections to ratings data in an attempt to compensate for systematic discrepancies attributed to different methods of data collection. See response error.

Cash-plus-Barter: a type of barter syndication in which the station pays the

APPENDIX B: GLOSSARY

syndicator cash, even though the program contains some advertising. See barter.

Cassandra: A syndicated research service of A.C. Nielsen, based on NSI data, which offers specialized reports on the performance of syndicated programs in markets across the country. Issued quarterly.

CATV (Community Antenna Television): an acronym for cable television, used in many early FCC proceedings.

CDC (Control Data Corporation): the computer manufacturer that owns Arbitron.

Census: a study in which every member of a population in interviewed or measured. Every 10 years, the federal government conducts a census of the U.S. population.

Channel Loyalty: a common phenomenon of aggregate audience behavior in which the audience for one program tends to be disproportionately represented in the audience for other programs on the same channel. See audience duplication, inheritance effects.

Circulation: the total number of unduplicated audience members exposed to a media vehicle (e.g., newspaper, station) over some specificed period of time. See cume, reach.

Clearance: the assurance given by a station that it will air a program feed by its affiliated network.

Cluster Sample: a type of probability sample in which aggregations of sampling units, called *clusters* (e.g., census tracts), are sampled at some stage in the process. See probablity sample.

CODE (Cable On-line Data Exchange): a service of A.C. Nielsen that maintains information on the stations and networks carried on all U.S. cable systems. See cable system.

Codes: the numbers or letters used to represent reponses in a survey instrument like a diary. Coding the responses allows computers to manipulate the data.

Coincidental: a type of telephone survey in which interviewers ask repondents what they are watching or listening to at the time of the call. Coincidentals, based on probability samples, often set the standard against with other ratings methods are judged.

COLRAM (Committee on Local Radio Audience Measurement): a committee of the NAB concerned with a range of local radio measurement issues.

COLTAM (Committee on Local Television Audience Measurement): a committee of the NAB concerned with a range of local television measurement issues.

COLTRAM (Committee on Local Television and Radio Audience Measurement): a committee of the NAB which, in 1985, was divided into COLRAM and COLTAM.

Condensed Market Report: a ratings report on a small or specialized market area that contains less information than a standard market report.

Confidence Interval: in probability sampling, it is the range of values around an estimated population value (e.g., a rating) with a given probability (i.e., confidence level) of encompassing the true population value.

Confidence Level: in probability sampling, it is a statement of the likelihood that a range of values (i.e., confidence interval) will include the true population value.

CMSA (Consolidated Metropolitan Statistical Area): a type of metropolitan area, designated by the U.S. Office of Management and Budget, often used by ratings companies to define a market's metro area.

Convenience Sample: a nonprobability sample, sometimes called an accidental sample, that is used because respondents are readily available or convenient.

Correlation: a statistic that measures the strength and direction of the relationship between two variables. It may range in value from +1.0 to –1.0, with 0 indicating no relationship.

CPP (Cost Per Point): a measure of how much it costs to buy the audience represented by one ratings point. The size of that audience, and therefore its cost, varies with the size of the market population on which the rating is based.

CPM (Cost Per Thousand): a measure of how much it costs to buy 1,000 audience members delivered by an ad. CPMs are commonly used to compare the cost efficiency of different advertising vehicles.

Cohort: a type of longitudinal survey design in which several independent samples are drawn from a population whose membership does not change over time. See longitudinal.

Counter programming: a programming strategy in which a station or network schedules material appealing to an audience other than the competition. Independents often counter program local news with entertainment.

Coverage: the potential audience for a given station or network, defined by the size of the population that is reached, or covered, by the signal.

Cross-sectional: a type of survey design in which one sample is drawn from the population at a single point in time. See longitudinal.

Cross-tabs: a technique of data analysis in which the responses to one item are paired with those of another item. Cross-tabs are useful in determining the audience duplication between two programs. See audience duplication.

Cume: short for cumulative audience, it is the size of the total unduplicated audience for a station over some specified period of time. When the cume is expressed as percent of the market population they are referred to as *cume ratings*. See circulation, reach.

Cume Duplication: the size of a station's audience that has also been in another station's audience, within some specified period of time. See exclusive cume.

Daypart: a specified period of time, usually defined by certain hours of the day and days of the week (e.g., weekdays vs. weekends), that is to summarize average audience size, or buy and sell advertising time.

Demographics: a category of variables often used to describe the composition of audiences. Common demographics include age, gender, education, occupation, and income.

DMA (Designated Market Area): the term used by A.C. Nielsen to describe specific market areas. Every county belongs to one, and only one, DMA. See ADI.

Diary: a paper booklet, distributed by ratings companies, in which audience members are asked to record their television or radio use, usually for one week.

DST (Differential Survey Treatment): special procedures used by a ratings company to improve response from segments of the population known to have

APPENDIX B: GLOSSARY

unusually low response rates. These may include additional interviews and incentives to cooperate.

Duplicated Audience: See audience duplication.

Early Fringe: in television, a daypart in late afternoon immediately prior the airing of local news programs.

Editing: the procedures used by a ratings company to check the accuracy and completeness of the data it collects. Editing may include techniques for clarifying or eliminating questionable data. See ascription, flip.

Effective Exposure: a concept in media planning stipulating that a certain amount of exposure to an advertising message is necessary before it is effective. Often used interchangeably with the term *effective frequency*. See frequency.

EMRC (Electronic Media Rating Council): an industry organization representative of advertising agencies and the electronic media that is responsible for accrediting the procedures used by ratings companies.

ESB (Effective Sample Base): the same as ESS.

ESF (Expanded Sample Frame): a procedure used by Arbitron to include in its sample frame households whose telephone numbers are unlisted. See sample frame.

ESS (Effective Sample Size): the size of a simple random sample needed to produce the same result as the sample actually used by the rating company. ESS is a convenience used for calculating confidence intervals.

Exclusive Cume Audience: the total size of the unduplicated audience that listens exclusively to one station within some specified period of time.

FCC (Federal Communications Commission): The independent regulatory agency, created in 1934, that has primary responsibility for the oversight of broadcasting and cable.

Flip: an editing procedure in which call letters erroneously entered in a diary are changed to those of the most plausible existing station. See editing.

Format: the style of programming offered by a radio station. Common formats include MOR (middle of the road), CHR (contemporary hits radio), and AOR (album-oriented rock).

Frame: See sample frame.

Frequency: in advertising, the number of times that an individual is exposed to a particular advertising message.

Frequency Distribution: a way representing the number of times different values of a variable occur within a sample or population.

Fringe: in television, dayparts just before (4–7 p.m. ET) and after prime time (11 p.m. on).

Geodemographics: a type of variable that categorizes audiences by combining geographic and demographic factors. For example, organizing audiences by zip codes with similar population age and income.

Geographic Weight: a mathematical adjustment to survey data used to correct the over- or under-representation of some geographic area.

Generic Entries: listening or viewing mentions in a diary that are ambiguously attributed to one particular station.

Grazing: the term describing the tendency of viewers to frequently change channels, a behavior that is presumably facilitated by remote control. See inertia.

Gross Impressions: the sum of all audiences exposed to an advertising schedule over a period of time. The number of gross impressions may exceed the size of the population since audience members may be duplicated. See GRP.

GRP (Gross Rating Point): the gross impressions of an advertising schedule expressed as a percentage of the population. GRPs are commonly used to describe to overall size or media weight of an advertising campaign.

Group Quarters: dormitories, barracks, nursing homes, prisons, and other living arrangements that do not qualify as households, and are, therefore, not measured by ratings companies.

Hammocking: a television programming strategy in which an unproven or weak show is scheduled between two popular programs in hopes that viewers will stay tuned thereby enhancing the rating of the middle program. See audience flow, inheritance effect.

Headend: the part of a cable system that receives TV signals from outside sources (e.g., off-the-air, satellite) and sends them through the wired distribution system. See cable system.

HDBA (High Density Black Area): a geographic area with a disproportionately large Black population, triggering differential survey treatment. See DST.

HDHA (High Density Hispanic Area): a geographic area with a disproportionately large Hispanic population, triggering differential survey treatment. See DST.

Home County: the county in which a station's city of license is located.

Home Market: the market area in which a station is located.

Home Station: any station licensed in a city within a given market area.

Hour-by-Hour Estimates: a section of a radio market report that estimates audience size in consecutive hours.

Household: an identifiable housing unit, such as an apartment or detached home, occupied by one or more persons. See group quarters.

Household Diary: a diary designed to record the viewing of any member of a household, or household visitor, who watches a particular television set. See diary, personal diary.

HPDV (Households Per Diary Value): the number of households in the population represented by a single diary kept by a sample household. Used to make audience projections. See projected audience.

HUT (Households Using Television): a term describing the total size of the audience, in households, at any point in time. Expressed as either the projected audience size, or as a percent of the total number of households.

Hypoing: any one of several illegal practices in which a station, or its agent, engages in an attempt to artificially inflate the station's rating during a measurement period. Also called *hyping*. See ratings distortion.

Independents: a commercial television station that does not maintain an affiliation with a broadcast network.

Inertia: a description of audience behavior that implies viewers are unlikely to change channels unless provoked by very unappealing programming. See grazing.

Inheritance Effect: a common phenomenon of television audience behavior, in which the audience for one program is disproportionately represented in the audience of the following program. Sometimes called *lead-in effects,* audience inheritance can be thought of as a special case of channel loyalty. See audience duplication, audience flow, channel loyalty.

Instantaneous Meter: See SIA

In-Tab: term describing the sample of households or persons actually used in tabulating or processing results.

Interview: a method of collecting data through oral questioning of a respondent, either in person, or over the telephone.

Interviewer: the researcher who conducts an interview.

Interviewer Bias: the problem of introducing systematic error or distortions in data collected in an interview, attributable to the appearance, manner, or reactions of the interviewer. See response error.

Late Fringe: in television, a daypart just after the late local news (11 p.m. ET).

Lead-In: the program that immediately precedes another on the same channel. The size and composition of a lead-in audience is an important determinant of program's rating. See inheritance effect.

Lead-In Effect: See inheritance effect.

Listening Mention: the basic unit of radio audience measurement, typically defined as a minimum of 5 minutes listening to a specific station.

Longitudinal: a type of survey designed to collect data over several points in time. See cross-sectional.

LOP (Least Objectionable Program): a popular theory of television audience behavior, from Paul Klein, that argues people primarily watch TV for reasons unrelated to content, and they choose the least objectionable programs. See passive audience.

Mean: a measure of central tendency determined by adding across cases, and dividing that total by the number of cases. See average, median, mode.

Measure: a procedure or device for quantifying objects (e.g., households, people) on variables of interest to the researcher.

Measurement: the process of assigning numbers to objects according to some rule of assignment.

Measurement Error: systematic bias or inaccuracy attributable to measurement procedures.

Mechanical Diary: a file or printout of the coded entries contained in a single diary. See codes, diary.

Median: a measure of central tendency defined as that point in a distribution where half the cases have higher values, and half have lower values. See average, mean, mode.

Meter: a measuring device used to record the on-off and channel tuning condition of a TV set. See SIA, peoplemeter.

Metro Area: the core metropolitan counties of a market area as defined by a ratings service. Metro generally correspond to MSAs.

Metro Mail: a company that supplies ratings companies with lists of names, addresses, and phone numbers from which samples are drawn.

Metro Rating: a program or station rating based on the behavior of those who live in the metro area of the market. See rating.

Metro Share: a program of station share based on the behavior of those who live in the Metro area of the market. See share.

Mode: a measure of central tendency defined as the value in a distribution that occurs most frequently. See average, mean, median.

Mortality: a problem of losing sample members over time, typically in longitudinal survey research.

MRA (Metro Rating Area): See metro.

MRS (Minimum Reporting Standard): the number of listening or viewing mentions necessary for a station or program to be included in a ratings report.

MSA (Metropolitan Statistical Area): an urban area designated by the Office of Management and Budget often used to by ratings companies to define their metro areas.

MSO (Multiple System Owner): a company owning more than one cable system.

Multi-Set Household: a television household with more than one working television set.

Multi-Stage Sample: a type of probability sample requiring more than one round of sampling. See cluster sample, probability sample.

NAB (National Association of Broadcasters): an industry organization representing the interests of commercial broadcasters.

NATPE (National Association of Television Program Executives): an industry organization of media professional responsible for television programming.

Narrowcasting: a programming strategy in which a station or network schedules content of the same type or appealing to the same subset of the audience. See block programming.

NCTA (National Cable Television Association): an industry organization representing the interests of cable systems.

Net Audience: See cume or reach.

Net Weekly Circulation: the cume or unduplicated audience using a station or network in a week. See cume.

Network: an organization that acquires or produces programming and distributes that programming, usually with national or regional advertising, to affiliated stations or cable systems.

Network Rating: the rating achieved by network or network program based on the entire U.S. population or the population within its coverage area.

Nielsen: a major supplier of national and local market television ratings.

Non ADI Station: a station whose home market is not within the ADI. See ADI.

Non-ADI Market: a one county market area created to accommodate a station whose home county is within an ADI, but outside the metro area. See ADI.

Non-DMA Station: a station whose home market is not within the DMA. See DMA.

Non-Probability Sample: a kind of sample in which every member of the population does not have a known probability of selection into the sample. See convenience sample, purposive sample, quota sample.

Non-Response: the problem of failing to obtain information from each person originally drawn into the sample.

Non-Response Error: biases or inaccuracies in survey data the result from non-response. See nonresponse.

Normal Distribution: a kind of frequency distribution that, when graphed, forms

APPENDIX B: GLOSSARY

a symmetrical, bell-shaped curve. Many statistical procedures are premised on the assumption that variables are normally distributed. See skew.

NSI (Nielsen Station Index): a division within the A.C. Nielsen company that issues a series of local television market ratings reports.

NSI Plus: a computerized service of NSI that allows customized analyses of audience flow, reach, and frequency.

NTI (Nielsen Television Index): a division of the A.C. Nielsen company that issues a series of national television network ratings.

Off-Network Programs: programs originally produced to air on a major broadcast network, now being sold in syndication.

Opportunistic Market: the buying and selling of network advertising time on short notice, as unforseen developments (e.g., cancellation, schedule changes) create opportunities. See scatter market, up-front market.

O & O (Owned & Operated): a broadcast station that is owned and operated by a major broadcast network.

Overnights: the label given to ratings, based on meters or coincidentals, that are available to clients the day after broadcast.

Oversample: deliberately drawing a sample larger than needed in-tab, to compensate for nonresponse, or to intensively study some subset of the sample.

Panel: a type of longitudinal survey design in which the same sample of individuals is studied over some period of time. Meters are placed in a panel of households. See cross-sectional, longitudinal, trend.

Passive Audience: a term given to viewers who are unselective about the content they watch. Passive audiences are thought to watch TV out of habit, tuning to almost anything if a preferred show is unavailable. See active audience, LOP.

Pay Cable: the programming services provided by a cable system for a monthly fee above and beyond that required for basic cable. Pay cable may include anyone of several "premium" services like HBO, Showtime, or The Disney Channel.

Peoplemeter: a device that electronically records the on-off and channel tuning condition of a TV set, and is capable of identifying who is in the room at the time of viewing. If viewer must enter that information by button pressing, the meter is called *active,* if the meter requires no effort from viewers, it is called *passive.*

Periodicity: a problem encountered in systematic sampling in which the sampling interval corresponds to some cyclical arrangement in the list.

Personal Diary: a small booklet in which a single person records his or her TV viewing or radio listening, usually for 1 week. Personal diaries are designed to accompany a person wherever media use occurs. See household diary.

Personal Interview: a method of data collection in which interviewer and respondent are face-to-face. See interview.

Placement Interview: an initial interview to secure the willingness of the respondent to keep a diary or receive a meter.

Planned No-Viewing Household: a household that, during a placement interview, indicates is will not be watching TV during the survey period.

Pocketpiece: the common name given to Nielsen's national TV ratings report.

Population: the total number of persons or households from which a probability sample is drawn. Membership in a population must be clearly defined, often by the geographic area in which a person lives.

PMSA (Primary Metropolitan Statistical Area): an urban area designated by the Office of Management and Budget that is often used in defining ratings areas.

PPDV (Persons Per Diary Value): the number of persons in a population represented by a single diary kept by a member of a ratings sample. PPDV is used to project an audience. See projected audience.

Preempt: an action, taken by an affiliate, in which programming fed by a network is replaced with programming scheduled by the station. Certain types of commercial time can also be "preempted" by advertisers willing to pay a premium for the spot.

Prime Time: a television daypart from 7 p.m. to 11 p.m. ET. Due to FCC regulations, broadcast networks typically feed programming only from 8 p.m. to 11 p.m. ET.

Probability Sample: a kind of sample in which every member of the population has an equal or known chance of being selected into the sample. Sometimes called *random samples*, probability samples allow statistical inferences about the accuracy of sample estimates. See confidence interval, confidence level, sampling error.

Processing Error: a source of inaccuracies in ratings reports attributable to problems mechanics of gathering and producing the data. See ascription, codes, editing.

Program Rating: the rating of a specific radio or television program.

Program Type: a category of programming usually based on similarities in program content. Nielsen identifies over 35 program types, used in summarizing program audiences.

Projectable: a quality describing a sample designed in such a way that audience projections may be made. See projected audience, probability sample.

Projected Audience: the total size of an audience estimated to exist in the population, based on sample information. See HPDV, PPDV, probability sample.

Psychographics: a category of variable that draws distinctions among people on the basis of their psychological characteristics, including opinions, interests, and attitudes.

PTAR (Prime Time Access Rule): an FCC regulation limiting the amount of network programming that affiliates in the top 50 markets can air between 7 p.m. and 11 p.m. ET.

PUR (Persons Using Radio): a term describing the total size of the radio audience at any point in time. See HUT, PUT.

Purposive Sample: a type of nonprobability sample, sometimes called a *judgment sample* sample, in which the researcher uses his or her knowledge of the population to "hand pick" areas or groups of respondents for research.

PUT (Person Using Television): a term describing the total size of the television audience, in persons, at any point in time. See HUT, PUR.

Qualitative Ratings: numerical summaries of the audience that not only describe how many watched or listened, but their reactions including enjoyment, interest, attentiveness, and information gained.

Qualitative Research: Any systematic investigation of the audience that does not depend on measurement and quantification. Examples include focus groups and participant observation. Sometimes used to describe any nonratings research, even if quantification is involved, as in "qualitative ratings."

Quota Sample: a type of nonprobability sample in which categories of respondents called *quotas* (e.g., males), are filled by interviewing respondents who are convenient. See nonprobability sample, probability sample.

RAB (Radio Advertising Bureau): an industry organization formed to promote advertising on radio.

RADAR (Radio's All Dimension Audience Research): a syndicated ratings service for radio network audiences offered by Statistical Research, Inc.

RADAR On Line (RAL): a service offering an on-line computer access to the RADAR database.

Random Digit Dialing (RDD): in telephone surveys, a technique for a creating probability sample by randomly generating telephone numbers. By using this method, all numbers including unlisted have an equal chance of being called.

Randon Sample: see probability sample

Rate of Response: the percentage of those originally drawn into the sample who provide useable information. See in-tab.

Rate Card: a list of how much a station will charge for its commercial spots. Rate cards are sometimes incorporated with ratings data in computer programs that manage station inventories.

Rating: in its simplest form, the percentage of persons or households tuned to a station or program out of the total market population.

Ratings Distortion: activity on the part of a broadcaster designed to alter the way audience members report their use of stations. See hypoing.

Reach: the total number of unduplicated persons or households included in the audience of a station or a commercial campaign over some specified period of time. Sometimes expressed as a percentage of the total market population. See cume, frequency.

Recycling: the extent to which listeners in one daypart also listen in another daypart. See audience duplication.

Relative Standard Error: a means of comparing the amount of sampling error in ratings data to the size of different ratings. It is the ratio of the standard error to the rating itself. See sampling error.

Relative Standard Error Thresholds: the size of a rating needed to have a relative standard error of either 25% or 50%. Often published in market reports as a means of judging ratings accuracy. See relative standard error.

Reliability: the extent to which a method of measurement yield consistent results over time.

Repeat Viewing: the extent to which the audience for one program is represented in the audience of other episodes of the series. See audience duplication.

Replication: A study repeating the procedures of an early study to assess the stability of results. In audience measurement, replications involve drawing subsamples from a parent sample to assess sampling error.

Respondent: a sample member who provides information in response to questions.

Response Error: inaccuracies in survey data attributable to the quality of responses, including lying, forgetting, or misinterpreting questions. See interviewer bias.

Roadblock: a strategy of scheduling commercial announcements in which an ad is run simultaneously on all three major networks to increase its reach.

Rolling Average: a ratings level based on the average of several successive samples. As new sample data become available, the oldest sample is dropped from the average. A rolling average is less susceptible to sampling error. See bounce.

ROS (Run of Schedule): a method of buying and scheduling ads in which the advertiser allows the station or network run commercials at the best time that happens to be available.

Sample: a subset of some population. See probability sample.

Sample Balancing: See sample weighting.

Sample Frame: a list of some population from with a probability sample is actually drawn.

Sample Weighting: the practice of assigning different mathematical weights to various subsets of the in-tab sample in an effort to correct for different response rates among those subsets. Each weight is the ratio of subset's size in the population to its size in the sample.

Sampling Distribution: the hypothetical frequency distribution of sample statistics that would result from repeated samplings of some population.

Sampling Error: inaccuracies in survey data attributable to "the luck of the draw" in creating a probability sample.

Sampling Rate: the ratio of sample size to population size.

Sampling Unit: the survey element (e.g., person or household), or aggregation of elements, considered for selection at some stage in the process of probability sampling.

Scatter Market: a period of time, just in advance of a given quarter of the year, during with advertisers by network time. See opportunistic market, up-front market.

Segmentation: the practice of dividing the total market into subsets, often related to the needs of a marketing plan or the programming preferences of the population. See target audience.

Sets-in-Use: the total number of sets turned on at a given point in time. As a measure of total audience size, it has become outdated since most households now have multiple sets. See HUT.

Share: in its simplest form, the percentage of persons or households tuned to a station or program out of all those using the using the medium at that time.

Share of Voice: a term used to express a particular brand's advertising presence relative to others in the same category. Often determined by each brand's total advertising expenditures.

SIA (Storage Instantaneous Audimeter): a latter version of Nielsen's original audimeter that allowed the company to retrieve electronically stored information over telephone lines.

APPENDIX B: GLOSSARY

Simple Random Sample: a one-stage probability sample in which every member of the population has an equal chance of selection. See probability sample.

SIP (Station Information Packet): forms on which broadcast stations report programs schedules and other information necessary to produce a local market ratings report.

Skew: a measure of the extent to which a frequency distribution departs for a normal, symmetrical shape. In common use, the extent to which some subset of population is disproportionately represented in the audience (e.g., "the audience skews old").

Spill: the extent to which nonmarket stations are viewed by local audiences, or local stations are viewed by audiences outside the market.

Spin-Off: a programming strategy in which the characters or locations of a popular program are used to create another television series.

Split Entry: an editing procedure used to reconcile diary entries that report listening to different stations in overlapping time periods.

SMSA (Standard Metropolitan Statistical Area): the former governmental designation of an urban area, once used by ratings companies to define local market areas. See MSA.

SRDS (Standard Rate and Data Service): a service that publishes the station rate cards and other information useful in buying commercial time. See rate card.

SRI (Statistical Research Incorporated): the company that published the RADAR radio network ratings and conducts other customized ratings research.

Standard Deviation: a measure of the variability in a frequency distribution.

Standard Error: the standard deviation of a sampling distribution. It is the statistic used to make statements about the accuracy of estimates based on sample information. See confidence interval, confidence level, relative standard error.

Station Rep: an organization that represents local stations to national and regional advertisers, selling the station's time and sometimes providing research information useful in programming.

Station Total Area: a Nielsen term meaning the total geographic area upon which total station audience estimates are based. The total area may include counties outside the NSI area.

Statistical Significance: the point at which results from a sample deviate so far from what could happen by chance that they are thought to reflect real differences or phenomena in the population. By convention, significance levels are usually set at .05 or lower, meaning a result could happen by chance only 5 times in 100. See confidence level.

Stratified Sample: a type of probability sample in which the population organized into homogeneous subsets or strata, after which a predetermined number of respondents is randomly selected for each strata. Stratified sampling can reduce the sampling error associated with simple random samples.

Stripped Programming: a programming practice in which television shows are scheduled at the same time on 5 consecutive weekdays. Stations often "strip" syndicated programs.

Superstation: an independent television station whose programming is widely carried on cable systems around the country. Superstations include WTBS, WGN, WWOR, WPIX, and KTTV.

Sweep: In television, a 4-week period of time during which ratings companies are collecting the audience information necessary to produce local market reports.

Syndicated: a standardized product sold to many clients. A syndicated program is available to stations in many different markets. A syndicated ratings report is also sold to many users.

Syndication: the marketing of syndicated programs.

Systematic Sample: a kind of probability sample in which a set interval is applied to a list of the population to identify elements included into the sample (e.g., picking every 10th name).

TALO (Total Audience Listing Output): An Arbitron radio ratings report listing all listening mentions within a specified geographic area.

Target Audience: any well defined subset of the total audience that an advertiser wants to reach with a commercial campaign, or a station wanted to reach with a particular kind of programming.

Telephone Recall: a type of survey in which a telephone interviewer asks the respondent what they listened to or watched in the recent past, often the preceding day. See coincidental.

Telephone Retrieval: the practice of recovering audience information recorded on a diary or questionnaire provided before hand by the ratings company.

Television Household: a common unit of analysis in ratings research, it is any household equipped with a working television set, excluding group quarters. See group quarters.

Tent-Poling: a programming strategy in which weak or untried programs are scheduled before and after a popular program in the hope that audience flow will enhance the ratings of the weaker adjacent programs. See inheritance effect.

Theory: a tentative explanation of how some phenomenon of interest works. Theories identify causes and effects which make them amenable to testing and falsification.

Tiering: the practice of marketing cable services to subscribers in groups or bundles of channels called *tiers*.

Time Buyers: anyone who buys time from the electronic media for purposes of running commercial announcements.

Time Period Averages: the size of a broadcast audience at an average point in time, within some specified period of time.

Torus: Arbitron's term for a geographic area surrounding, but not including, a smaller geographic area. Hence the MSA + the ADI Torus = ADI.

Total Audience: all those who tune to a program for at least 5 minutes. Essentially, it is the cumulative audience for a long program or miniseries.

Trend Analysis: a type of longitudinal survey design in which results from repeated independent samplings are compared over time.

TSA (Total Survey Area): Arbitron's designation for all the included in a market report, including those outside the ADI.

TSL (Time Spent Listening): a cumulative measure of the average amount of time an audience spends listening to a station within a daypart.

Turnover: the ratio of a station's cumulative audience to its average quarter hour within a daypart.

TVB (Television Advertising Bureau): an industry organization formed to promote advertising on broadcast television.

TVQ: a ratings system that assesses the familiarity and likability of personalities and programs.

UHF (Ultra High Frequency): a class television stations assigned to broadcast on channels 14 through 80.

Unduplicated Audience: the number of different persons or households in an audience over some specified period of time.

Unit of Analysis: the element or entity about which a researcher collects information. In ratings the unit of analysis is usually a person or household.

Universe: See Population.

Unweighted In-Tab: the actual number of individuals in different demographic groups who have returned usable information to the ratings company.

Unwired Networks: organizations that acquire commercial time from stations around the country and package that time for sale in the up-front market, without offering a common schedule of programming.

Up-Front Market: a period of time several months in advance of the new fall television during which networks, barter syndicators, and major advertisers agree to the sale of large blocks of commercial time in the coming season.

Validity: the extent to which a method of measurement accurately quantifies the attribute it is supposed to measure.

Variable: any well-defined attribute or characteristic that varies from person to person, or thing to thing. See demographic.

VCR (Video Cassette Recorder): an appliance used for recording and playing video cassette tapes, now in a majority of U.S. households.

Viewing Mention: the unit of television diary measurement, typically defined as at least 5 minutes of viewing a particular station or program. See listening mention.

VHF (Very High Frequency): a class of television stations assigned to broadcast on channels 2 through 13.

VPVH (Viewers Per Viewing Household): the estimated number of people, usually by demographic category, in each household tuned to a particular source.

Weighted In-Tab: the number of individuals in different demographic groups who would have provided usable information if response rates were equivalent. See sample weighting.

Weighting: the process of assigning mathematical weigths in an attempt to correct over or under-representation of some groups in the unweighted in-tab sample. See sample weighting.

Working Women: a commonly reported subset of the television audience, including adult females who work outside the home for 30 or more hours a week.

Zapping: the practice of using a remote control device to avoid commercials or program content by rapidly changing channels. Often used interchangeably with zipping.

Zipping: the practice of using the fast forward on a VCR to speed through unwanted commercials or program content. Often used interchangeably with zapping

APPENDIX C:
ADI MARKET RANKING (1989–1990)*

1989-1990 TELEVISION MARKET RANKINGS

are based on Arbitron estimates of U.S. Television Households as of January 1, 1990. Markets in parentheses ( ) have no ADI of their own.

The television audience estimates for the non-ADI home stations are included in their corresponding market report.

The following dates are for the 1989-1990 survey periods.

October 1989	March 1990
Sept. 27 - Oct. 24	Feb. 28 - March 27
November 1989	**May 1990**
November 1 - 28	April 25 - May 22
January 1990	**July 1990**
January 3 - 30	July 11 - August 7
February 1990	
Jan. 31 - Feb. 27	

The Arbitron Meter Service provides daily and biweekly reports for:

Atlanta	*Denver	Miami
Boston	Detroit	New York
Chicago	Houston	Philadelphia
Cleveland	Los Angeles	San Francisco
Dallas-Ft. Worth		Washington, DC

*Single-source reports are produced for this market.

The Arbitron Weekly Television Service provides ADI and TSA demographic estimates for Pittsburgh.

New Season	Second Season
Sept. 27 - Oct. 3	January 3 - 9
October 4 - 10	January 10 - 16
October 11 - 17	January 17 - 23
November 1 - 7	Jan. 31 - Feb. 6
November 8 - 14	February 7 - 13

ADI MARKET RANKINGS

RANK	MARKET	ADI TV HH
1	New York (Kingston, Poughkeepsie & Bridgeport, CT)	7,043,900
2	Los Angeles (Corona & San Bernardino-Fontanta)	4,939,400
3	Chicago (La Salle)	3,124,800
4	Philadelphia (Allentown, Reading, Vineland, Wildwood & Wilmington, DE)	2,704,400
5	San Francisco-Oakland-San Jose (Santa Rosa & Vallejo)	2,200,700
6	Boston (Derry, Manchester & Worcester)	2,105,800
7	Dallas-Ft. Worth	1,728,900
8	Detroit	1,723,500
9	Washington, DC	1,701,700
10	Houston	1,453,200
11	Cleveland (Akron, Canton & Sandusky)	1,442,000
12	Atlanta (Rome, GA)	1,374,900
13	Minneapolis-St. Paul	1,333,200
14	Tampa-St. Petersburg (Lakeland & Sarasota)	1,323,000
15	Seattle-Tacoma (Bellingham)	1,296,500
16	Miami (Ft. Lauderdale)	1,288,800
17	Pittsburgh	1,165,500
18	St. Louis (Mt. Vernon, IL)	1,121,500
19	Denver (Steamboat Springs)	1,047,400
20	Phoenix (Kingman & Prescott, AZ)	1,004,100
21	Sacramento-Stockton	995,800
22	Baltimore	937,400
23	Hartford-New Haven (New London)	901,000
24	San Diego	867,500
25	Orlando-Daytona Beach-Melbourne (Leesburg, FL)	862,800
26	Indianapolis (Marion, IN)	851,000
27	Portland, OR	804,100
28	Milwaukee (Kenosha, WI)	762,100
29	Kansas City (Lawrence)	755,900
30	Cincinnati	744,600
31	Charlotte (Hickory)	729,500
32	Nashville (Cookeville)	720,600
33	Raleigh-Durham (Fayetteville, NC & Goldsboro, NC)	693,600
34	Columbus, OH (Chillicothe & Mansfield)	685,500
35	Greenville-Spartanburg-Asheville (Anderson, SC & Toccoa, GA)	646,900
36	New Orleans	640,800
37	Grand Rapids-Kalamazoo-Battle Creek (Muskegon, MI)	623,200
38	Buffalo	611,600

(continued)

---

*(Reprinted with permission from Arbitron Company)

ADI MARKET RANKINGS (continued)

RANK	MARKET	ADI TV HH
39	Memphis	605,400
40	Oklahoma City	600,200
41	Salt Lake City	598,200
42	Norfolk-Portsmouth-Newport News-Hampton	575,400
43	San Antonio (Kerrville)	573,200
44	Providence-New Bedford (Vineyard Haven)	556,900
45	Harrisburg-York-Lancaster-Lebanon	552,300
46	Louisville	546,400
47	Birmingham (Gadsden, AL)	525,300
48	Charleston-Huntington	518,800
49	Greensboro-Winston Salem-High Point (Burlington, NC)	515,100
50	West Palm Beach-Ft. Pierce-Vero Beach	513,700
51	Albuquerque (Santa Fe & Hobbs, NM)	503,900
52	Dayton (Richmond, IN)	501,300
53	Albany-Schenectady-Troy	489,500
54	Wilkes Barre-Scranton	489,200
55	Mobile-Pensacola (Ft. Walton Beach, FL)	467,900
56	Jacksonville	464,100
57	Little Rock	461,500
58	Tulsa (Bartlesville)	460,200
59	Flint-Saginaw-Bay City	452,900
60	Richmond	442,800
61	Wichita-Hutchinson	434,700
62	Fresno-Visalia (Hanford)	427,400
63	Toledo	409,700
64	Knoxville (Crossville)	407,200
65	Shreveport-Texarkana	406,000
66	Des Moines	382,400
67	Green Bay-Appleton (Suring, WI)	381,600
68	Syracuse	371,500
69	Roanoke-Lynchburg	368,200
70	Lexington (Beattyville, Danville & Hazard)	366,900
71	Austin, TX	357,800
72	Rochester, NY	353,000
73	Omaha	351,300
74	Portland-Poland Spring	336,000
75	Springfield-Decatur-Champaign	335,000
76	Paducah-Cape Girardeau-Harrisburg-Marion	328,100
77	Spokane	320,500
78	Davenport-Rock Island-Moline: Quad City (Burlington, IA)	315,500
79	Tucson	307,900
80	Huntsville-Decatur-Florence	307,000
81	Cedar Rapids-Waterloo-Dubuque	305,800
82	Columbia, SC	299,200
83	Springfield, MO	296,500

RANK	MARKET	ADI TV HH
84	Chattanooga (Cleveland, TN)	295,100
85	South Bend-Elkhart	294,500
86	Jackson, MS (Natchez)	292,900
87	Bristol-Kingsport-Johnson City: Tri Cities (Greenville, TN)	282,900
88	Johnstown-Altoona	280,400
89	Youngstown	277,000
90	Madison	274,700
91	Las Vegas	270,600
92	Burlington-Plattsburgh (Hartford, VT-Hanover, NH)	265,100
93	Evansville (Madisonville, KY)	263,900
94	Baton Rouge	253,600
95	Lincoln-Hastings-Kearney	253,400
96	Ft. Myers-Naples	247,500
97	Waco-Temple-Bryan	247,200
98	Springfield, MA	244,700
99	Colorado Springs-Pueblo	237,900
100	Sioux Falls-Mitchell	235,900
101	Ft. Wayne (Angola)	230,900
102	Savannah	229,800
103	Lansing (Ann Arbor)	225,000
104	El Paso	224,300
105	Greenville-New Bern-Washington	224,200
106	Charleston, SC	221,300
107	Montgomery-Selma	218,900
108	Fargo	218,000
109	Salinas-Monterey	210,600
110	Augusta	209,300
111	Peoria-Bloomington	208,200
112	Santa Barbara-Santa Maria-San Luis Obispo (Oxnard)	205,500
113	McAllen-Brownsville: LRGV	198,400
114	Lafayette, LA	195,900
115	Ft. Smith	194,400
116	Tallahassee-Thomasville	188,500
117	Reno	187,300
118	Amarillo	185,100
119	Columbus, GA (Opelika)	184,700
120	Monroe-El Dorado	183,200
121	Eugene	182,400
122	Corpus Christi	174,000
122	Macon	174,000
124	Tyler-Longview-Jacksonville	170,600
125	Duluth-Superior	170,200
126	Yakima	169,300
127	La Crosse-Eau Claire	165,700

APPENDIX C: ADI MARKET RANKING

RANK	MARKET	ADI TV HH
128	Beaumont-Port Arthur	164,500
129	Columbus-Tupelo (West Point)	164,400
130	Wausau-Rhinelander	163,000
131	Traverse City-Cadillac	162,100
132	Wichita Falls-Lawton	161,900
133	Terre Haute	161,800
134	Binghamton	158,600
135	Boise	156,700
136	Rockford	156,300
137	Wheeling-Steubenville	156,100
138	Sioux City	155,800
139	Erie	154,000
140	Florence-Myrtle Beach	153,800
141	Chico-Redding	153,100
142	Bakersfield	152,100
143	Topeka	151,300
144	Bluefield-Beckley-Oak Hill	150,200
145	Minot-Bismarck-Dickinson-Glendive	144,600
146	Odessa-Midland	143,300
147	Wilmington	143,300
148	Joplin-Pittsburg	143,000
149	Rochester-Mason City-Austin	142,000
150	Lubbock	137,500
151	Albany, GA (Valdosta & Cordele)	137,200
152	Medford	133,500
153	Columbia-Jefferson City	133,400
154	Quincy-Hannibal	121,200
155	Bangor	116,300
156	Abilene-Sweetwater	115,800
157	Clarksburg-Weston	111,500
158	Dothan	104,500
159	Utica	103,600
160	Idaho Falls-Pocatello	101,600
161	Salisbury	93,400
162	Laurel-Hattiesburg	91,000
163	Alexandria, LA	88,200
164	Gainesville (Ocala)	87,700
165	Rapid City	86,400
166	Billings-Hardin	85,000
167	Elmira	84,100
168	Greenwood-Greenville	82,200
169	Panama City	82,000
170	Watertown-Carthage	81,900
171	Missoula	79,200
172	Lake Charles	75,800
173	Ardmore-Ada	73,800
174	Jonesboro	73,100
175	Palm Springs	67,900
176	Meridian	67,300
177	Biloxi-Gulfport-Pascagoula	66,700
178	El Centro-Yuma	65,200
179	Great Falls	64,800
180	Grand Junction-Durango	64,700
181	Jackson, TN	58,800
182	Marquette	55,300
183	Tuscaloosa	54,900
184	Eureka	52,100
185	Butte	48,600
186	San Angelo	48,200
187	St. Joseph	47,500
188	Anniston	45,400
189	Cheyenne-Scottsbluff (Sterling)	44,400
190	Bowling Green (Campbellsville)	44,200
191	Lafayette, IN	43,300
192	Hagerstown	42,100
193	Casper-Riverton	42,000
194	Lima	41,200
195	Charlottesville	40,900
196	Parkersburg	36,200
197	Laredo	36,100
198	Harrisonburg	35,900
199	Zanesville	32,000
200	Twin Falls	30,000
201	Presque Isle	29,300
202	Flagstaff	28,700
203	Ottumwa-Kirksville (Wapello, IA)	28,600
204	Bend	26,300
205	Victoria	26,100
206	Mankato	23,100
207	Helena	19,400
208	North Platte	18,000
209	Alpena	15,500

REFERENCES*

Advertising Research Foundation. (1954). *Recommended standards for radio and television audience size measurements*. New York: Author.

Adams, W.J., Eastman, S.T., Horney, L.J., & Popovich, M.N. (1983). The cancelation and manipulation of network television prime-time programs. *Journal of Communication, 33*(1), 10–27.

Agostini, J.M. (1961). How to estimate unduplicated audiences. *Journal of Advertising Research, 1,* 11–14.

Agostini, J.M. (1962). Analysis of magazine accumulative audience. *Journal of Advertising Research, 2,* 24–27.

Agostino, D. (1977). *The cable subscriber's viewing of public television: A comparison of public television use between broadcast viewers and cable subscribers within selected markets*. Bloomington, IN: Institute for Communication Research.

Agostino, D. (1980). Cable television's impact on the audience of public television. *Journal of Broadcasting, 24,* 347–363.

Agostino, D., & Zenaty, J. (1980). *Home vcr owner's use of television and public television: Viewing, recording & playback*. Washington, DC: Corporation for Public Broadcasting.

Allen, C. (1965). Photographing the TV audience. *Journal of Advertising Research, 5,* 2–8.

Allen, R. (1981). The reliability and stability of television exposure. *Communication Research, 8,* 233–256.

American Research Bureau. (1947, May). *Washington DC market report*. Beltsville, MD: Author.

Anderson, J.A. (1987). *Communication research: Methods and issues*. New York: McGraw-Hill.

Anderson, J.A., & Meyer, T.P. (1988). *Mediated communication: A social action perspective*. Newbury Park, CA: Sage.

---

*This reference section includes all works cited in the text as well as other relevant readings.

REFERENCES

Arbitron. (1987). *Arbitron year-round.* New York: Author.
Arbitron. (1989). *Milwaukee-Racine market report: Spring 1989.* New York: Author.
Atkin, C., Greenberg, B., Korzenny, F.K. & McDermott, S. (1978). Selective exposure to televised violence. *Journal of Broadcasting, 22,* 47–61.
Atkin, D., & Litman, B. (1986). Network TV programming: Economics, audiences, and the ratings game, 1971–1986. *Journal of Communication, 36*(3), 32–51.
Austin, B.A. (1989). *Immediate seating: A look at movie audiences.* Belmont, CA: Wadsworth.
Babbie, E.R. (1983). *The practice of social research* (3rd ed.) New York: Wadsworth.
Babrow, A.S., & Swanson, D.L. (1988). Disentangling antecedents of audience exposure levels: Extending expectancy-value analyses of gratifications sought from television news. *Communication Monographs, 55,* 1–21.
Banks, M. (1981). *A history of broadcast audience research in the United States, 1920–1980 with an emphasis on the rating services.* Unpublished doctoral dissertation, University of Tenessee, Knoxville, TN.
Banks, S. (1980). Children's television viewing behavior. *Journal of Marketing, 44,* 48–55.
Barnes, B.E., & Thompson, L.M. (1988). The impact of audience information sources on media evolution. *Journal of Advertising Research, 28,* RC9-RC14.
Baron, R. (1988). If it's on computer paper, it must be right. *Journal of Media Planning, 2,* 32–34.
Barwise, T.P. (1986). Repeat-viewing of prime-time television series, *Journal of Advertising Research, 26,* 9–14.
Barwise, T.P., & Ehrenberg, A.S.C. (1984). The reach of TV channels. *International Journal of Research in Marketing, 1,* 34–49.
Barwise, T.P, & Ehrenberg, A.S.C. (1988). *Television and its audience.* London: Sage.
Barwise, T.P., Ehrenberg, A.S.C., & Goodhardt, G.J. (1979). Audience appreciation and audience size. *Journal of Market Research Society, 21,* 269–289.
Barwise, T.P., & Ehrenberg, A.S.C., & Goodhart, G.J. (1982). Glued to the box?: Patterns of TV repeat-viewing. *Journal of Communication, 32*(4), 22–29.
Bechtel, R.K. Achelpohl, C., & Akers, R. (1972). Correlation between observed behavior and questionnaire responses on television viewing. In E.A. Rubinstein, G.A. Comstock, & J.P. Murray (Eds.), *Television and social behavior: Vol. 4. Television in day-to-day life: Patterns of use* (pp. 274–344). Washington, DC: U.S. Government Printing Office.
Becknell, J.C. (1961). The influence of newspaper tune-in advertising on the size of a TV show's audience. *Journal of Advertising Research, 1,* 23–26.
Beebe, J.H. (1977). The institutional structure and program choices in television markets. *Quarterly Journal of Economics, 91,* 15–37.
Becker, L.B., & Schoenback, K. (Eds.). (1989). *Audience responses to media diversification: Coping with plenty.* Hillsdale, NJ: Lawrence Erlbaum Associates.
Benz, W. (1988). The advertiser. *Gannett Center Journal, 2*(3), 90–93.
Besen, S.M., Krattenmaker, T.G., Metzger, A.R., & Woodbury, J.R. (1984). *Misregulating television: Network dominance and the FCC.* Chicago: University of Chicago Press.

Besen, S.M., & Mitchell, B.M. (1976). Watergate and television: An economic analysis. *Communication Research, 3,* 243–260.
Beville, H.M., Jr. (1988). *Audience ratings: Radio, television, cable* (rev. ed). Hillsdale, NJ: Lawrence Erlbaum Associates.
Blum, R.A., & Lindheim, R.D. (1987). *Primetime: Network television programming.* Boston: Focal Press.
Blumler, J.G. (1979). The role of theory in uses and gratifications studies. *Communication Research, 6,* 9–36.
Blumler, J.G., Gurevitch, M., & Katz, E. (1985). Reaching out: A future for gratifications research. In K. Rosengren, L. Wenner & P. Palmgreen (Eds.), *Media gratifications research: Current perspectives* (pp. 255–273). Beverly Hills: Sage.
Boemer, M.L. (1987). Correlating lead-in show ratings with local television news ratings. *Journal of Broadcasting & Electronic Media, 31,* 89–94.
Bogart, L. (1972). *The age of television.* New York: Frederick Ungar.
Bogart, L. (1988). Research as an instrument of power. *Gannett Center Journal, 2*(3), 1–16.
Bortz, P.I. (1986). *Great expectations: A television manager's guide to the future.* Washington, DC: National Association of Broadcasters.
Bower, R.T. (1973). *Television and the public.* New York: Holt, Rinehart & Winston.
Bower, R.T. (1985). *The changing television audience in America.* New York: Columbia University Press.
Bowman, G.W., & Farley, J. (1972). TV viewing: Application of a formal choice model. *Applied Economics, 4,* 245–259.
Boyer, P. (1988). Bewitched, bothered, and bewildered. *Gannett Center Journal, 2*(3), 17–26.
Brotman, S.N. (1988). *Broadcasters can negotiate anything.* Washington, DC: National Association of Broadcasters.
Bruno, A.V. (1973). The network factor in TV viewing. *Journal of Advertising Research, 13,* 33–39.
Bryant, J., & Zillmann, D. (1984). Using television to alleviate boredom and stress: Selective exposure as a function of induced excitational states. *Journal of Broadcasting, 28,* 1–20.
Bryant, J., & Zillmann, D. (Eds.). (1986). *Perspectives on media effects.* Hillsdale, NJ: Lawrence Erlbaum Associates.
Burnett, L. (1989). *1989 media costs and coverage.* Chicago, IL: Leo Burnett U.S.A.
Byrne, B. (1988). Barter syndicators. *Gannett Center Journal, 2*(3), 75–78.
Cabletelevision Advertising Bureau. (1983). *Cable audience methodology study.* New York: Author.
Cabletelevision Advertising Bureau. (1990). *Cable TV facts.* New York: Author.
Cannon, H.M. (1983). Reach and frequency estimates for specialized target markets. *Journal of Advertising Research, 23,* 45–50.
Cannon, H., & Merz, G.R. (1980). A new role for psychographics in media selection. *Journal of Advertising, 9*(2), 33–36.
Cantril, H., & Allport, G.W. (1935). *The psychology of radio.* New York: Harper & Brothers.
CBS. (1937). *Radio in 1937.* New York: Author.

REFERENCES

Chaffee, S. (1980). Mass media effects: New research perspectives. In D.C. Wilhoit & H. DeBock (Eds.), *Mass communication review yearbook* (pp. 77–108). Beverly Hills, CA: Sage.

Chandon, J.L. (1976). *A comparative study of media exposure models*. Unpublished doctoral dissertation, Northwestern University, Evanston, IL.

Chappell, M.N., & Hooper, C.E. (1944). *Radio audience measurement*. New York: Stephen Daye.

Christ, W., & Medoff, N. (1984). Affective state and selective exposure to and use of television. *Journal of Broadcasting, 28,* 51–63.

Clift, C., & Greer, A. (Eds.). (1981). *Broadcasting programming: The current perspective*. Washington, DC: University Press of America.

Cohen, E. E. (1989). *A model of radio listener choice*. Unpublished doctoral dissertation, Michigan State University. East Lansing, MI.

Collins, J., Reagan, J., & Abel, J. (1983). Predicting cable subscribership: Local factors. *Journal of Broadcasting, 27,* 177–183.

Comstock, G., Chaffee, S., Katzman, N., McCombs, M., & Roberts, D. (1978). *Television and human behavior*. New York: Columbia University Press.

Comstock, G. (1989). *The evolution of American television*. Newbury Park, CA: Sage.

CONTAM. (1989). *Nielsen people meter review: 1988–1989*. Westfield, NJ: Statistical Research, Inc.

Converse, T. (Speaker). (1974, May 2). *Magazine* [Television documentary] New York: CBS, Inc.

Cook, F. (1988, January). Peoplemeters in the USA: An historical and methodological prespective. *Admap,* 32–35.

Corporation for Public Broadcasting. (1980). *Proceedings of the 1980 technical conference on qualitative television ratings: Final report*. Washington, DC: Author.

Csikszentmihalyi, M., & Kubey, R. (1981). Television and the rest of life: A systematic comparison of subjective experience. *Public Opinion Quarterly, 45,* 317–328.

Darmon, R. (1976). Determinants of TV viewing. *Journal of Advertising Research, 16,* 17–20.

Dimling, J. (1988). A. C. Nielsen: The "gold standard." *Gannett Center Journal, 2*(3), 63–69.

Dominick, J.R., & Fletcher, J.E. (1985). *Broadcasting research methods*. Boston: Allyn & Bacon.

Ducey, R., Krugman, D., & Eckrich, D. (1983). Predicting market segments in the cable industry: The basic and pay subscribers. *Journal of Broadcasting, 27,* 155–161.

Duncan, J.H. (semi-annual, Spring and Fall). *American radio*. Indianapolis, IN: Duncan's American Radio, Inc.

Eastman, S.T., Head, S.W., & Klein, L. (1989). *Broadcast/cable programming: Strategies and practices* (3rd ed). Belmont, CA: Wadsworth.

Ehrenberg, A.S.C. (1968). The factor analytic search for program types. *Journal of Advertising Research, 8,* 55–63.

Ehrenberg, A.S.C. (1982). *A primer in data reduction*. London & New York: Wiley.

Ehrenberg, A.S.C., & Wakshlag, J. (1987). Repeat-viewing with people meters. *Journal of Advertising Research, 27,* 9–13.

Elliot, P. (1974). Uses and gratifications research: A critique and a sociological alternative. In J.G. Blumler & E. Katz (Eds.), *The uses of mass communications: Current perspectives on gratifications research* (pp. 249–268). Beverly Hills: Sage.

Ettema, J.S., & Whitney, C.D. (Eds.). (1982). *Individuals in mass media organizations: Creativity and constraint.* Beverly Hills: Sage.

Federal Communications Commission. (1979). *Inquiry into the economic relationship between television broadcasting and cable television* (71 F.C.C. 2d 241). Washington, DC: U.S. Government Printing Office.

Fisher, F.M., McGowan, J.J., & Evans, D.S. (1980). The audience-revenue relationship for local television stations. *Bell Journal of Economics, 11,* 694–708.

Fletcher, A.D., & Bower, T.A. (1988). *Fundamentals of advertising research* (3rd ed.). Belmont, CA: Wadsworth.

Fletcher, J.E. (Ed.). (1981). *Handbook of radio and TV broadcasting: Research procedures in audience, program and revenues.* New York: Van Nostrand Reinhold.

Fletcher, J.E. (1985). *Squeezing profits out of ratings: A manual for radio managers, sales managers and programmers.* Washington, DC: National Association of Broadcasters.

Fletcher, J.E. (1987). *Music and program research.* Washington, DC: National Association of Broadcasters.

Fletcher, J.E. (1988). *Broadcast research definitions.* Washington, DC: National Association of Broadcasters.

Foote, J. S. (1988). Ratings decline of presidential television. *Journal of Broadcasting and Electronic Media, 32,* 225–230.

Fournier, G.M., & Martin, D.L. (1983). Does government-restricted entry produce market power? New evidence from the market for television advertising. *Bell Journal of Economics, 14,* 44–56.

Fowler, M. S., & Brenner, D.L. (1982). A marketplace approach to broadcast regulation. *Texas Law Review, 60,* 207–257.

Frank, R.E., Becknell, J., & Clokey, J. (1971). Television program types. *Journal of Marketing Research, 11,* 204–211.

Frank, R.E., & Greenberg, M.G. (1980). *The public's use of television.* Beverly Hills: Sage.

Fratrik, M. R. (1989, April). *The television audience-revenue relationship revisited.* Paper presented at the meeting of the Broadcast Education Association, Las Vegas, NV.

Gantz, W., & Eastman, S.T. (1983). Viewer uses of promotional media to find out about television programs. *Journal of Broadcasting, 27,* 269–277.

Gantz, W., & Razazahoori, A. (1982). The impact of television schedule changes on audience viewing behaviors. *Journalism Quarterly, 59,* 265–272.

Gans, H. (1980). The audience for television-and in television research. In S.B. Witney & R.P. Abeles (Eds.), *Television and social behavior: Beyond violence and children* (pp. 55–81). Hillsdale, NJ: Lawrence Erlbaum Associates.

REFERENCES

Garrison, G. (1939). Wayne University. *Journal of Applied Psychology, 23,* 204–205.

Gensch, D.H. (1969, May). A computer simulation model for selecting advertising schedules. *Journal of Marketing Research, 6,* 203–214.

Gensch, D.H., & Ranganathan, B. (1974). Evaluation of television program content for the purpose of promotional segmentation. *Journal of Marketing Research, 11,* 390–398.

Gensch, D.H., & Shaman, P. (1980). Models of competitive ratings. *Journal of Marketing Research, 17,* 307–315.

Gerbner, G., Gross, L., Morgan, M., & Signorielli, N. (1986). Living with television: The dynamics of the cultivation process. In J. Bryant & D. Zillmann (Eds.), *Perspectives on media effects* (pp. 17–40). Hillsdale, NJ: Lawrence Erlbaum Associates.

Gitlin, T. (1983). *Inside prime time.* New York: Pantheon.

Goodhardt, G.J. (1966). The constant in duplicated television viewing between and within channels. *Nature, 212,* 1616.

Goodhardt, G.J., & Ehrenberg, A.S.C. (1969). Duplication of viewing between and within channels. *Journal of Marketing Research, 6,* 169–178.

Goodhardt, G.J., Ehrenberg, A.S.C., & Collins, M.A. (1987). *The television audience: Patterns of viewing* (2nd ed.). Westmead, UK: Gower.

Glasser, G.J., & Metzger, G.D. (1989, December). *SRI/CONTAM review of the Nielsen people meter: The process and the results.* Paper presented at the eight annual Advertising Research Foundation Electronic Media Workshop, New York.

Grant, A. E. (1989). *Exploring patterns of television viewing: A media system dependency perspective.* Unpublished doctoral dissertation, University of Southern California, Los Angeles, CA.

Greenberg, E., & Barnett, H.J. (1971). TV program diversity—New evidence and old theories. *American Economic Review, 61,* 89–93.

Greenberg, B., Dervin, B., & Dominick, J. (1968). Do people watch 'television' or 'programs'?: A measurement problem. *Journal of Broadcasting, 12,* 367–376.

Hall, R. W. (1988). *Media math: Basic techniques of media evaluation.* Lincoln, IL: NTC Business Books.

Headen, R., Klompmaker, J., & Rust, R. (1979). The duplication of viewing law and television media schedule evaluation. *Journal of Marketing Research, 16,* 333–340.

Headen, R.S., Klompmaker, J.E., & Teel, J.E. (1977). Predicting audience exposure to spot TV advertising schedules. *Journal of Marketing Research, 14,* 1–9.

Headen, R.S., Klompmaker, J.E., & Teel, J.E. (1979). Predicting network TV viewing patterns. *Journal of Advertising Research, 19,* 49–54.

Heeter, C., & Greenberg, B. (1985). Cable and program choice. In D. Zillmann & J. Bryant (Eds.), *Selective exposure to communication* (pp. 203–224). Hillsdale, NJ: Lawrence Erlbaum Associates.

Heeter, C., & Greenberg, B.S. (1985). Profiling the zappers. *Journal of Advertising Research, 25*(2), 15–19.

Heeter, C., & Greenberg, B.S. (1988). *Cable-viewing.* Norwood, NJ: Ablex.

Heighton, E.J., & Cunningham, D.R. (1984). *Advertising in the broadcasting media* (2nd ed.). Belmont, CA: Wadsworth.

Henriksen, F. (1985). A new model of the duplication of television viewing: A behaviorist approach. *Journal of Broadcasting & Electronic Media, 29,* 135–145.

Herzog, H. (1944). What do we really know about daytime serial listeners? In P.J. Lazersfeld & F.N. Stanton (Eds.), *Radio research 1942–1943* (pp. 23–36). New York: Duell, Sloan & Pearce.

Hiber, J. (1987). *Winning radio research: Turning research into ratings and revenues.* Washington, DC: National Association of Broadcasters.

Hill, D., & Dyer, J. (1981). Extent of diversion to newscasts from distant stations by cable viewers. *Journalism Quarterly, 58,* 552–555.

Hirsch, P. (1980). An organizational perspective on television (Aided and abetted by models from economics, marketing, and the humanities). In S.B. Withey & R.P Abeles (Eds.), *Television and social behavior* (pp. 83–102). Hillsdale, NJ: Lawrence Erlbaum Associates.

Hogarth, R.M., & Reder, M.W. (Eds.). (1986). *Rational choice: The contrast between economics and psychology.* Chicago, IL: University of Chicago Press.

Horen, J.H. (1980). Scheduling of network television programs. *Management Science, 26,* 354–370.

Hotelling, H. (1929). Stability in competition. *Economic Journal, 34,* 41–57.

Howard, H.H., & Kievman, M.S. (1983). *Radio and TV programming.* New York: Macmillan.

Israel, H., & Robinson, J. (1972). Demographic characteristics of viewers of television violence and news prgrams. In E.A. Rubinstein, G.A. Comstock, & J.P. Murray (Eds.), *Television and social behavior: Vol. 4. Television in day-to-day life: Patterns of use* (pp. 87–128). Washington, DC: U.S. Government Printing Office.

Jaffe, M. (1985, January 25). Towards better standards for post-analysis of spot television GRP delivery. *Television/Radio Age,* pp. 23–25.

Jeffres, L.W. (1978). Cable TV and viewer selectivity. *Journal of Broadcasting, 22,* 167–177.

Jeffres, L.W. (1986). *Mass media processes and effects.* Prospect Heights, IL: Waveland.

Jhally, S., & Livant, B. (1986). Watching as working: The valorization of audience consciousness. *Journal of Communication, 36*(3), 124–143.

Kaplan, B.M. (1985). Zapping—The real issue is communication. *Journal of Advertising Research, 25* (2), 9–12.

Kaplan, S.J. (1978). The impact of cable television services on the use of competing media. *Journal of Broadcasting, 22,* 155–165.

Kaatz, R.B. (1982). *Cable: An advertiser's guide to the new electronic media.* Chicago: Crain Books.

Katz, E., Blumler, J.G., & Gurevitch, M. (1974). Utilization of mass communication by the individual. In J.G. Blumler & E. Katz (Eds.) *The uses of mass communications: Current perspectives on gratifications research* (pp. 19–32) Beverly Hills: Sage.

Katz, E., Gurvitch, M., & Haas, H. (1973). On the use of mass media for important things. *American Sociological Review, 38*(2), 164–181.

REFERENCES

Killion, K.C. (1987). Using peoplemeter information. *Journal of Media Planning, 2*(2), 47–52.

Kirsch, A.D., & Banks, S. (1962). Program types defined by factor analysis. *Journal of Advertising Research, 2,* 29–31.

Klapper, J. (1960). *The effects of mass communication.* Glencoe, IL: The Free Press.

Klein, P. (1971, January). The men who run TV aren't stupid. . . . *New York,* pp. 20–29.

Krugman, D.M. (1985). Evaluating the audiences of the new media. *Journal of Advertising, 14*(4), 21–27.

Krugman, H.E. (1972). Why three exposures may be enough. *Journal of Advertising Research, 12,* 11–14.

Kubey, R. W. (1986). Television use in everyday life: Coping with unstructured time. *Journal of Communication, 36*(3), 108–123.

Kubey, R., & Csikszentmihalyi, M. (1990). *Television and the quality of life: How viewing shapes everyday experience.* Hillsdale, NJ: Lawrence Erlbaum Associates.

LaRose, R., & Atkin, D. (1988). Satisfaction, demographic, and media environment predictors of cable subscription. *Journal of Broadcasting & Electronic Media, 32,* 403–413.

Lavine, J.M., & Wackman, D.B. (1988). *Managing media organizations: Effective leadership of the media.* New York: Longman.

Lazarsfeld, P.F., & Stanton, F.N. (Eds.). (1941). *Radio research.* New York: Duell, Sloan & Pearce.

Lazer, W. (1987). *Handbook of demographics for marketing and advertising.* Lexington, MA: Lexington Books.

Leckenby, J.D., & Rice, M.D. (1985). A beta binomial network TV exposure model using limited data. *Journal of Advertising, 3,* 25–31.

LeDuc, D. R. (1973). *Cable television and the FCC: A crisis in media control.* Philadelphia: Temple University Press.

LeDuc, D. R. (1987). *Beyond broadcasting: Patterns in policy and law.* New York: Longman.

Lehmann, D.R. (1971). Television show preference: Application of a choice model. *Journal of Marketing Research, 8,* 47–55.

Levin, H.G. (1980). *Fact and fancy in television regulation: An economic study of policy alternatives.* New York: Russell Sage.

Levy, M.R. (1978). The audience experience with television news. *Journalism Monographs, 55.*

Levy, M. R. (Ed.). (1989). *The VCR age: Home video and mass communication.* Newbury Park, CA: Sage.

Levy, M.R., & Fink, E.L. (1984). Home video recorders and the transience of television broadcasts. *Journal of Communication, 34*(2), 56–71.

Levy, M., & Windahl, S. (1984). Audience activity and gratifications: A conceptual clarification and exploration. *Communication Research, 11,* 51–78.

Lichty, L., & Topping, M. (Eds.). (1975). *American broadcasting: A source book on the history of radio and television.* New York: Hastings House.

Lindolf, T.R. (Ed.). (1987). *Natural audiences: Qualitative research on media uses and effects.* Norwood, NJ: Ablex.

Litman, B.R. (1979). Predicting TV ratings for theatrical movies. *Journalism Quarterly, 56,* 591–594.

Little, J.D.C. & Lodish, L.M. (1969). A media planning calculus. *Operations Research, 1,* 1–35.

LoSciuto, L.A. (1972). A national inventory of television viewing behavior. In E.A. Rubinstein, G.A. Comstock, & J.P. Murray (Eds.), *Television and social behavior: Vol. 4. Television in day-to-day life: Patterns of use* (pp. 33–86). Washington, DC: U.S. Government Printing Office.

Lowery, S., & DeFleur, M.L. (1987). *Milestones in mass communication research: Media effects* (2nd ed.). New York: Longman.

Lu, D., & Kiewit, D.A. (1987). Passive peoplemeters: A first step. *Journal of Advertising Research, 27*(3), 9–14.

Lull, J. (1980). The social uses of television. *Human Communication Research, 6,* 197–209.

Lull, J. (1982). How families select televisions programs: A mass observational study. *Journal of Broadcasting, 26,* 801–812.

Lull, J. (Ed.). (1988). *World families watch television.* Newbury Park, CA: Sage.

Lumley, F. H. (1934). *Measurement in radio.* Columbus, OH: The Ohio State University.

MacFarland, D.T. (1990). *Contemporary radio programming strategies.* Hillsdale, NJ: Lawrence Erlbaum Associates.

McCombs, M.E., & Shaw, D.L. (1972). The agenda-setting function of the mass media. *Public Opinion Quarterly 36,* 176–87.

McDonald, D.G., & Reese, S.D. (1987). Television news and audience selectivity. *Journalism Quarterly, 64,* 763–768.

McDonald, D.G., & Schechter, R. (1988). Audience role in the evolution of fictional television content. *Journal of Broadcasting & Electronic Media, 32,* 61–71.

McLeod, J.M., & McDonald, D.G. (1985). Beyond simple exposure: Media orientations and their impact on political processes. *Communication Research, 12,* 3–33.

McPhee, W.N. (1963). *Formal theories of mass behavior.* New York: The Free Press.

McQuail, D. (1987). *Mass communication theory: An introduction.* London: Sage.

McQuail, D., & Gurevitch, M. (1974). Explaining audience behavior: Three approaches considered. In J.G. Blumler & E. Katz (Eds.), *The uses of mass communications: Current perspectives on gratifications research* (pp. 287–302). Beverly Hills: Sage.

Meehan, E.R. (1984). Ratings and the institutional approach: A third answer to the commodity question. *Critical Studies in Mass Communication, 1,* 216–225.

Metheringham, R.A. (1964). Measuring the net cumulative coverage of a print campaign. *Journal of Advertising Research, 4,* 23–28.

Metzger, G. (1984, February). Current audience measurement is doing the job; Meters at local level should be viewed cautiously. *Television/Radio Age,* pp 46–47.

Miller, P.V. (1987, May). *Measuring TV viewing in studies of television effects.* Paper presented at the meeting of the International Communication Association, Montreal.

REFERENCES

Miller, P. (1988). I am single source. *Gannett Center Journal 2*(3), 27–34.
Morley, D. (1986). *Family television: Cultural power and domestic leisure.* London: Comedia.
Myrick, H.K., & Keegan, C. (1981). *Boston (WGBH) field testing of a qualitative television rating system for public broadcasting.* Washington DC: Corporation for Public Broadcasting.
Naples, M.J. (1979). *The effective frequency: The relationship between frequency and advertsising effectiveness.* New York: Association of National Advertisers.
Neuman, W.R. (1982). Television and American culture: The mass medium and the pluralist audience. *Public Opinion Quarterly, 46,* 471–487.
Newcomb, H.M., & Alley, R.S. (1983). *The producer's medium.* New York: Oxford University Press.
Newcomb, H.M., & Hirsch, P.M. (1984). Television as a cultural forum: Implications for research. In W. Rowland & B. Watkins (Eds.), *Interpreting television* (pp. 58–73). Beverly Hills, CA: Sage.
Nielsen, A. C. (1988). Television ratings and the public interest. In J. Powell & W. Gair (Eds.), *Public interest and the business of broadcasting: The broadcast industry looks at itself* (pp. 61–63). New York: Quorum Books.
Niven, H. (1960). Who in the family selects the TV program? *Journalism Quarterly, 37,* 110–111.
Noam, E. (Ed.). (1985). *Video media competition: Regulation, economics, and technology.* New York: Columbia Univeristy Press.
Noll, R.G., Peck, M.G., & McGowan, J.J. (1973). *Economic aspects of television regulation.* Washington, DC: Brookings Institution.
Ogburn, W.F. (1933). The influence of invention and discovery. In W.F. Ogburn (Ed.), *Recent social trends* (pp. 153–156). New York: McGraw-Hill.
Owen, B.M. (1975). *Economics and freedom of expression: Media structure and the first amendment.* Cambridge, MA: Ballinger.
Owen, B.M., Beebe, J., & Manning, W. (1974). *Television economics.* Lexington, MA: D.C. Heath.
Owen, B.M., & Wildman, S.W. (1991). *Video economics.* Cambridge, MA: Harvard University Press.
Palmgreen, P., Wenner, L.A., & Rayburn, J.D. (1981). Gratification discrepancies and news program choice. *Communication Research, 8,* 451–478.
Park, R.E. (1970). *Potential impact of cable growth on television broadcasting* (R-587-FF). Santa Monica, CA: Rand Corporation.
Park, R.E. (1979). *Audience diversion due to cable television: Statistical analysis of new data.* R-2403-FCC. Santa Monica, CA: Rand Corporation.
Parkman, A.M. (1982). The effect of television station ownership on local news ratings. *Review of Economics and Statistics, 64,* 289–295.
Perse, E.M. (1986). Soap opera viewing patterns of college students and cultivation. *Journal of Broadcasting and Electronic Media, 30,* 175–193.
Peterson, R. (1972). Psychographics and media exposure. *Journal of Advertising Research, 12,* 17–20.
Philport, J. (1980). The psychology of viewer program evaluation. In *Proceedings of the 1980 technical conference on qualitative ratings.* Washington, DC: Corporation for Public Broadcasting.

Poltrack, D. (1983). *Television marketing: Network, local, and cable.* New York: McGraw-Hill.
Poltrack, D. (1988). The "big 3" networks. *Gannett Center Journal, 2*(3), 53–62.
Radio Advertising Bureau. (1990). *Radio facts for advertisers 1990.* New York: Author.
Rao, V.R. (1975). Taxonomy of television programs based on viewing behavior. *Journal of Marketing Research, 12,* 335–358.
Reagan, J. (1984). Effects of cable television on news use. *Journalism Quarterly, 61,* 317–324.
Robinson, J. (1972). Television's immpact on everyday life: Some cross-national evidence. In E.A. Rubinstein, G.A. Comstock, & J.P. Murray (Eds.), *Television and social behavior: Vol. 4. Television in day-to-day life: Patterns of use* (pp. 410–431). Washington, DC: U.S. Government Printing Office.
Robinson, J.P. (1977). *How Americans used their time in 1965.* New York: Praeger.
Robinson, J.P., & Levy, M.R. (1986). *The main source: Learning from television news.* Beverly Hills, CA: Sage.
Rosenberg, M.J., & Hovland, C.I. (1960). Cognitive, affective, and behavioral components of attitudes. In M.J. Rosenberg, C.I. Hovland, W.J. McGuire, R.P. Abelson, & J.W. Brehm (Eds.), *Attitude organization and change: An analysis of consistency among attitude components* (pp. 1–16). New Haven: Yale University Press.
Rothenberg, J. (1962). Consumer sovereignty and the economics of TV programming. *Studies in Public Communication, 4,* 23–36.
Routt, E., McGrath, J.B., & Weiss, F.A. (1978). *The radio format conundrum.* New York: Hastings House.
Rowland, W. (1983). *The politics of TV violence: Policy uses of communication research.* Beverly Hills: Sage.
Rubens, W.S. (1978). A guide to TV ratings. *Journal of Advertising Research, 18,* 11–18.
Rubens, W.S. (1984). High-tech audience measurement for new-tech audiences. *Critical Studies in Mass Communication, 1,* 195–205.
Rubens, W.S. (1989). A personal history of TV ratings, 1929 to 1989 and beyond. *Feedback, 30*(4), 3–15.
Rubin, A.M. (1984). Ritualized and instrumental television viewing. *Journal of Communication, 34*(3), 67–77.
Rubin, A.M., & Perse, E.M. (1987). Audience activity and soap opera involvement. *Human Communication Research, 14,* 246–268.
Rubin, A.M., & Perse, E.M. (1987). Audience activity and television news gratifications. *Communication Research, 14,* 58–84.
Rubin, A.M., & Rubin, R.B. (1982). Contextual age and television use. *Human Communication Research, 8,* 228–244.
Rust, R.T. (1986). *Advertising media models: A practical guide.* Lexington, MA: Lexington Books.
Rust, R.T., & Alpert, M.I. (1984). An audience flow model of television viewing choice. *Marketing Science, 3*(2), 113–124.
Rust, R.T., & Donthu, N. (1988). A programming and positioning strategy for cable television networks. *Journal of Advertising, 17,* 6–13.

REFERENCES

Rust, R.T., & Klompmaker, J. E. (1981). Improving the estimation procedure for the beta binomial TV exposure model. *Journal of Marketing Research, 18,* 442–448.

Rust, R.T., Klompmaker, J.E., & Headen, R.S. (1981). A comparative study of television duplication models. *Journal of Advertising, 21,* 42–46.

Rust, R.T., & Leone, R.P. (1984). The mixed media dirichlet multinomial distribution: A model for evaluating television-magazine advertising schedules. *Journal of Marketing Research, 24,* 84–99.

Sabavala, D.J., & Morrison, D.G. (1977). A model of TV show loyalty. *Journal of Advertising Research, 17,* 35–43.

Sabavala, D.J., & Morrison, D.G. (1981). A nonstationary model of binary choice applied to media exposure. *Management Science, 27,* 637–57.

Salomon, G., & Cohen, A. (1978). On the meaning and validity of televison viewing. *Human Communication Research, 4,* 265–270.

Salvaggio, J.L., & Bryant, J. (Eds.). (1989). *Media use in the information age: Emerging patterns of adoption and consumer use.* Hillsdale, NJ: Lawrence Erlbaum Associates.

Schramm, W., Lyle, J., & Parker, E.B. (1961). *Television in the lives of our children.* Stanford, CA: Stanford University Press.

Schudson, M. (1984). *Advertising, the uneasy persuasion: Its dubious impact on American society.* New York: Basic Books.

Sears, D.O., & Freedman, J.L. (1972). Selective exposure to information: A critical review. In W. Schramm & D. Roberts (Eds.). *The process and effects of mass communication* (pp. 209–234). Urbana, IL: University of Illinois Press.

Schroder, K. (1987). Convergence of antagonistic traditions? The case of audience research. *European Journal of Communication, 2,* 7–31.

Sherman, B.L. (1987). *Telecommunications management: The broadcast and cable industries.* New York: McGraw-Hill.

Sieber, R. (1988). Cable networks. *Gannett Center Journal, 2*(3), 70–74.

Sims, J. (1988). AGB: The ratings innovator. *Gannett Center Journal, 2*(3), 85–89.

Singer, J.L., Singer, D.G., & Rapaczynski, W.S. (1984). Family patterns and television viewing as predictors of children's beliefs and aggression. *Journal of Communication, 34*(3), 73–89.

Sissors, J. Z., & Bumba, L. (1989). *Advertising media planning* (3rd ed.). Lincolnwood, IL: NTC Business Books.

Smythe, D. (1981). *Dependancy road: Communications, capitalism, consciousness, and Canada.* Norwood, NJ: Ablex.

Soong, R. (1988). The statistical reliability of people meter ratings. *Journal of Advertising Research, 28,* 50–56.

Sparkes, V. (1983). Public perception of and reaction to multi-channel cable television service. *Journal of Broadcasting, 27,* 163–175.

Spaulding, J. W. (1963). 1928: Radio becomes a mass advertising medium. *Journal of Broadcasting, 7,* 31–44.

Spence, M.A., & Owen, B.M. (1977). Television programming, monopolistic competition and welfare. *Quarterly Journal of Economics, 91,* 103–26.

Stanford, S.W. (1984). Predicting favorite TV program gratifications from general orientations. *Communication Research, 11,* 419–436.

Stanton, F.N. (1935). *Critique of present methods and a new plan for studying listening behavior.* Unpublished doctoral dissertation, The Ohio State University, Columbus, OH.

Steeves, H., & Bostain, L. (1982). A comparison of cooperation levels of diary and questionnaire respondents. *Journalism Quarterly, 4,* 610–616.

Steiner, G.A. (1963). *The people look at television.* New York: Alfred A. Knopf.

Steiner, G.A. (1966). The people look at commercials: A study of audience behavior. *Journal of Business, 39,* 272–304.

Steiner, P.O. (1952). Program patterns and preferences, and the workability of competition in radio broadcasting. *Quarterly Journal of Economics, 66,* 194–223.

Swanson, C.I. (1967). The frequency structure of television and magazines. *Journal of Advertising Research, 7,* 3–7.

Sterling, C.H., & Kittross, J.M. (1990). *Stay tuned: A concise history of american broadcasting* (2nd ed.). Belmont, CA: Wadsworth.

Statistical Research, Inc. (1975). *How good is the television diary technique?* Prepared for the National Association of Broadcasters. Washington, DC: Author.

Sudman, S., & Bradburn, N. (1982). *Asking questions: A practical guide to questionnaire design.* San Francisco: Jossey-Bass.

Television Audience Assessment. (1983a). *The audience rates television.* Boston, MA: Author.

Television Audience Assessment. (1983b). *The multichannel environment.* Boston, MA: Author.

Television Bureau of Advertising. (1990). *Trends in television.* New York: Author.

Tiedge, J.T., & Ksobiech, K.J. (1986). The "lead-in" strategy for prime-time: Does it increase the audience? *Journal of Communication, 36*(3), 64–76.

Tiedge, J.T., & Ksobiech, K.J. (1987). Counterprogramming primetime network television. *Journal of Broadcasting & Electronic Media, 31,* 41–55.

Turow, J. (1984). *Media industries.* New York: Longman.

Urban, C.D. (1984). Factors influencing media consumption: A survey of the literature. In B.M. Compaine (Ed.), *Understanding new media: Trends and issues in electronic distribution of information* (pp. 213–282). Cambridge, MA: Ballinger.

Vogel, H.L. (1986). *Entertainment industry economics: A guide for financial analysis.* Cambridge: Cambridge University Press.

Wakshlag, J., Agostino, D., Terry, H., Driscoll, P., & Ramsey, B. (1983). Television news viewing and network affiliation change. *Journal of Broadcasting, 27,* 53–68.

Wakshlag, J., Day, K., & Zillmann, D. (1981). Selective exposure to educational television programs as a function of differently paced humorous inserts. *Journal of Educational Psychology, 73,* 27–32.

Wakshlag, J., & Greenberg, B. (1979). Programming strategies and the popularity of television programs for children. *Human Communication Research, 6,* 58–68.

Wakshlag, J., Reitz, R., & Zillmann, D. (1982). Selective exposure to and acquisition of information from educational television programs as a function of appeal

and tempo of background music. *Journal of Educational Psychology, 74,* 666–677.
Wakshlag, J., Vial, V.K., & Tamborini, R. (1983). Selecting crime drama and apprehension about crime. *Human communication Research, 10,* 227–242.
Walker, J. R. (1988). Inheritance effects in the new media environment. *Journal of Broadcasting & Electronic Media, 32,* 391–401.
Wand, B. (1968). Television viewing and family choice differences. *Public Opinion Quarterly, 32,* 84–94.
Warner, C. (1986). *Broadcast and cable selling.* Belmont, CA: Wadsworth.
Waterman, D. (1986). The failure of cultural programming on cable TV: An economic interpretation. *Journal of Communication, 36*(3), 92–107.
Waterman, D. (1988, October). *Narrowcasting on cable television: A program choice model.* Paper presented at the Telecommunications Policy Research Conference, Airlie VA.
Webster, J.G. (1982). *The impact of cable and pay cable on local station audiences.* Washington, DC: National Association of Broadcasters.
Webster, J.G. (1983). The impact of cable and pay cable television on local station audiences. *Journal of Broadcasting, 27,* 119–126.
Webster, J.G. (1983). *Audience research.* Washington, DC: National Association of Broadcasters.
Webster, J.G. (1984, April). Peoplemeters. In *Research & Planning: Information for management.* Washington, DC: National Association of Broadcasters.
Webster, J.G. (1984). Cable television's impact on audience for local news. *Journalism Quarterly, 61,* 419–422.
Webster, J.G. (1985). Program audience duplication: A study of television inheritance effects. *Journal of Broadcasting & Electronic Media, 29,* 121–133.
Webster, J.G. (1986). Audience behavior in the new media environment. *Journal of Communication, 36*(3), 77–91.
Webster, J.G. (1989). Assessing exposure to the new media. In J. Salvaggio & J. Bryant (Eds.) *Media use in the information age: Emerging patterns of adoption and consumer use* (pp. 3–19). Hillsdale, NJ: Lawrence Erlbaum Associates.
Webster, J.G. (1989). Television audience behavior: Patterns of exposure in the new media environment. In J. Salvaggio & J. Bryant (Eds.) *Media use in the information age: Emerging patterns of adoption and consumer use* (pp. 197–216). Hillsdale, NJ: Lawrence Erlbaum Associates.
Webster, J.G. (1990). The role of audience ratings in communications policy. *Communications and the Law, 12*(2), 59–72.
Webster, J.G., & Coscarelli, W. (1979). The relative appeal to children of adult versus children's television programming. *Journal of Broadcasting, 23,* 437–451.
Webster, J.G., & Newton, G. D. (1988). Structural determinants of the television news audience. *Journal of Broadcasting & Electronic Media, 32,* 381–389.
Webster, J.G., & Wakshlag, J. (1982). The impact of group viewing on patterns of television program choice. *Journal of Broadcasting, 26,* 445–455.
Webster, J.G., & Wakshlag, J. (1983). A theory of television program choice. *Communication Research, 10,* 430–446.

Webster, J.G., & Wakshlag, J. (1985). Measuring exposure to television. In D. Zillmann & J. Bryant (Eds.), *Selective exposure to communication* (pp. 35–62). Hillsdale, NJ: Lawrence Erlbaum Associates.

Weibull, L. (1985). Structural factors in gratifications research. In K.E. Rosengren, L.A. Wenner, & P. Palmgreen (Eds.), *Media gratifications research: Current perspectives* (pp. 123–148). Beverly Hills: Sage.

Wells, W.D. (1969). The rise and fall of television program types. *Journal of Advertising Research, 9,* 21–27.

Wells, W.D. (1975). Psychographics: A critical review. *Journal of Marketing Research, 12,* 196–213.

White, K.J. (1977). Television market shares, station characteristics and viewer choice. *Communication Research, 4,* 415–434.

The why's and wherefores of syndex II. (1988, May 23). *Broadcasting,* pp. 58–59.

White, B.C., & Satterthwaite, N.D. (1989). *But first these messages. . . The selling of broadcast advertising.* Boston: Allyn and Bacon.

Wildman, S.S., & Owen, B.M. (1985). Program competition, diversity, and multichannel bundling in the new video industry. In E. Noam (Ed.), *Video media competition: Regulation, economics, and technology* (pp. 244–273). New York: Columbia University Press.

Wildman, S.S., & Siwek, S. E. (1988). *International trade in films and television programs.* Cambridge: Ballinger.

Wimmer, R.D., & Dominick, J.R. (1991). *Mass media research: An introduction* (3rd ed.). Belmont, CA: Wadsworth.

Wirth, M. O., & Bloch, H. (1985). The broadcasters: The future role of local stations and the three networks. In E. Noam (Ed.), *Video media competition: Regulation, economics, and technology* (pp. 121–137). New York: Columbia University Press.

Wirth, M.O., & Wollert, J.A. (1984). The effects of market structure on local television news pricing. *Journal of Broadcasting, 28,* 215–224.

Wober, J.M. (1988). *The use and abuse of television: A social psychological analysis of the changing screen.* Hillsdale, NJ: Lawrence Erlbaum Associates.

Wober, J.M., & Gunter, B. (1986). Television audience research at Britain's Independent Broadcasting Authority, 1974–1984. *Journal of Broadcasting and Electronic Media, 30,* 15–31.

Wulfemeyer, K.T. (1983). The interests and preferences of audiences for local television news. *Journalism Quarterly, 60,* 323–328.

Zeigler, S.K., & Howard, H. (1990). *Broadcast advertising* (3rd ed.) Ames, IA: Iowa State University Press.

Zenaty, J. (1988). The advertising agency. *Gannett Center Journal, 2*(3), 79–84.

Zillmann, D., & Bryant, J. (Eds.). (1985). *Selective exposure to communication.* Hillsdale, NJ: Lawrence Erlbaum Associates.

Zillmann, D., Hezel, R.T., & Medoff, N.J. (1980). The effect of affective states on selective exposure to televised entertainment fare. *Journal of Applied Social Psychology, 10,* 323–339.

AUTHOR INDEX

A

Agostini, J.M., 233
Allport, G., 43
Alpert, M.I., 212
Atkin, C., 214

B

Baron, R., 233
Barwise, T.P., 158, 230, 231, 236
Beebe, J., 57, 159
Beville, H.M., 67, 95
Blumler, J.G., 161–162
Bower, R.T., 163
Bryant, J., 161

C

Cantril, H., 43
Chandon, J.L., 235
Chappell, M.N., 71, 72
Collins, M.A., 172, 233, 236

E

Ehrenberg, A.S.C., 158, 172, 230, 231, 233, 236
Elliott, P., 178

F

Foote, J.S., 206–207

G

Garrison, G., 77–78
Gerbner, G., 54–55
Goodhardt, G.J., 172, 233–236
Gensch, D.H., 212
Gurevitch, M., 161–162

H

Headen, R.S., 235
Henriksen, F., 234, 235
Hooper, C., 71, 72
Horen, J.H., 212
Hovland, C.I., 159

I

Israel, H., 55

J

Jaffe, M., 211

K

Katz, E., 161–162
Klein, P., 161, 164, 209
Klompmaker, J.E., 235
Ksobiech, K.J., 214, 228

L

Lazarsfeld, P., 43
Litman, B., 213, 214
Lumley, F.H., 71

M

Manning, W., 57, 159
McDonald, D.G., 214
McPhee, W.N., 147
Metheringham, R.A., 233
Morgan, M., 55

N

Newton, G.D., 213
Nielsen, Jr., A. C., 56–57

O

Ogburn, W.F., 42
Owen, B.M., 57, 159

P

Park, R., 59
Parkman, A.M., 57–58, 214
Poltrack, D., 7, 199

R

Robinson, J., 55
Rosenberg, M.J., 159
Rust, R.T., 212, 213, 233, 235

S

Schechter, R., 214
Shaman, P., 212
Signorielli, N., 55
Siwek, S.E., 51
Spauling, J.W., 68
Stanton, F., 42, 67
Steiner, G., 151, 161, 163
Steiner, P., 159

T

Tiedge, J.T., 21, 228

V

Vogel, H.L., 45

W

Walker, J.R., 214, 228
Webster, J.G., 213, 235
Wildman, S.S., 51

Z

Zillmann, D., 161

SUBJECT INDEX

A

ABC (American Broadcasting Company), 10, 11, 121, 234
Accreditation, Electronic Media Rating Council, 141, 238, 239, 242
A. C. Nielsen Company, 61, 80, 124–133, 238–240
   cable television, 102, 115
   designated market area (DMA), 12, 125
   diaries, 102, 106
   history of, 72, 73, 76–77, 78, 80, 81, 82, 107
   peoplemeters, 80, 107, 108, 109, 115, 240
   radio index (NRI), 76, 82
   sampling, 87–88
   station index(NSI), 76, 82, 124, 130, 138, 239–240
   syndicated programming, 135–137, 240
   television index (NTI), 76, 80, 116, 120–123, 137, 189–190, 221–223, 238–239
Action for Children's Television (ACT), 53
Advertising and advertisers, 3–23, 196
   attitudes toward, 4
   cable television, 4–5, 8, 14, 202
   commercials, audiences, 100–101, 202, 239
   as consumers of ratings products, 119, 138–139, 140
   cost factors, 6–7, 12, 22–23, 45–46
   CPM/CPP, 23, 46, 48, 147, 186, 192–193, 199, 201–202, 239
   dayparts, concentration across, 47
   historical perspectives, 5, 69–70
   product use data, 138–139
   radio, 4–5, 10–11, 17–18, 47–48, 70, 71, 81
   remote control devices and, 106, 177
   revenues, 4–5, 47
   seasonal patterns, 7–8
   spot advertising, 4, 11, 13, 47
   supply/demand, 45–46
   syndication and, 11, 14–16, 37–38
   toll broadcasting, 68
Advertising agencies, 12–13, 15, 70, 72, 119, 177
Advertising Research Foundation (ARF), 101–102
Affiliates, network
   commercial time, 48
   cost of rating data, 140
   early fringe, 30–31
   market penetration, 170–171
   programming, 33, 167
   radio, 114
Age factors, 8, 32, 73, 121, 134, 146, 157, 162, 188, 196–198, 199–200, 203–204, 229
   adolescents, 8, 28–29, 162, 196
   young adults, 8, 105
   see also Children

281

American Association of Advertising Agencies, 70, 72
American Medical Association, 53
American Research Bureau, 74, 78–79
 also see Arbitron
AM radio, 10–11, 27, 132, 134, 166, 205, 243
Antitrust law, 58, 59
Arbitron Information on Demand (AID), 138–139, 241
Arbitron Ratings Company, 124–133, 240–241
 area of dominant influence (ADI), 12, 124–125, 130, 131, 188, 200, 261–263
 cable television, 58
 diaries, 102, 103, 104
 history, 79, 80
 methodology, 86–87, 102, 114
 minority groups, 97
 peoplemeters, 107, 241
 radio, 28, 29, 81, 102, 124, 131–133, 216, 240, 241
 television, 82, 124–131, 135, 240–241
 total survey area (TSA), 125, 131
Area of dominant influence (ADI), 12, 124–125, 130, 131, 188, 200, 261–263
Arnold, P, 71
Association of National Advertisers, 70, 72
Average quarter hour (AQH), 124, 134, 190–192, 196, 199, 217, 218, 242
Audience duplication, 171–173, 182, 223–228, 239
 inheritance effect, 38, 171–172, 226–227, 228, 234
 lead-in programming, 38, 135, 138, 171, 213, 214, 226–227
 mathematical modeling, 233–236
Audience loyalty, 19–21, 34, 61–62, 172–174, 227, 230, 232, 234
 radio, 173, 182–183, 230
Audimeters, 76, 79, 106
Audit Bureau of Circulation, 68
Audits of Great Britain (AGB), 80, 108, 109
Automated Measurement of Lineups (AMOL), 115

B

Barter syndication, 4, 14–16, 37–38, 167
Berra, Y., 29
Behavioral factors, 21

cumulative measures *vs* gross measures, 228, 229
 inheritance effect, 38, 171–172, 226–227, 228, 234
 radio measures, 34
 see also Preference analysis; Social psychology
Birch Radio, 80, 81, 110, 131–132, 134, 242
Birch, T., 81
Block programming, 38
Brand Cumulative Report, 222, 223
Broadcast Rating Council, *see* Electronic Media Rating Council
Bureau of Applied Social Research, 43

C

Cable Advertising Bureau, 202
Cable television, 4, 8–10, 53, 165, 174, 188
 advertising, 4–5, 8, 14, 202
 competition with broadcast, 47, 52, 58, 59–60, 157, 165, 169
 demographics, 240
 diaries, 102, 106
 litigation, 58, 60–61
 local markets, 14, 56, 58, 59–60, 114–115, 116, 151, 167, 241
 market penetration, 8–10, 165–166, 169, 170, 181, 202
 networks, 8–10, 47, 52, 57, 59–60, 157, 165, 169, 188, 195, 228, 240
 Nielsen Television Index (NTI), 76, 80, 116, 120–123, 137, 189–190, 221–223, 238–239
 prime time, 169
 subscription behavior, 174–175
Call out research, 28
CBS (Columbia Broadcasting System), 69, 70, 73, 78, 121
Census Bureau, 97
Chappell, M.N., 73
Children, 7, 53, 55, 61, 68, 162, 163–164
 adolescents, 8, 28–29, 162, 196
 diaries, 106
 meters, 108–109
Clark, M., 71–72
Commercials, *see* Advertising; Spot advertising
Competition
 antitrust law, 58, 59

SUBJECT INDEX

cable *vs* broadcast television, 47, 52, 58, 59–60, 157, 165, 169
  programming and, general, 168–169, 182
Computer applications
  audience size and composition software, 207
  databases, 81, 115, 119, 124, 137, 138–140, 149–150, 215, 241, 243
  diaries, 74, 112, 113
  meters, 108, 109–110
  for radio ratings, 28, 37, 124, 134, 239, 240, 241, 242
  for television ratings, 37, 130, 131, 239, 240, 241
Confidence intervals, 93
CONTAM, 94
Control Data Corporation, 79
  see also Arbitron
Cooper, R., 79
Cooperative Analysis of Broadcasting (CAB), 70–72, 73, 82, 84
Copyright, 52, 58, 60–61
  Copyright Royalty Tribunal, 52, 60
Corporation for Public Broadcasting, 33, 101
Cost factors, 43–44, 50, 196
  advertising, 6–7, 12, 22–23, 45–46
  CPM/CPP, 23, 46, 48, 147, 186, 192–193, 199, 201–202
  dayparts and, 39
  meters, 82, 107, 140–141
  programming, 39–41, 169
  ratings data, 70, 140–141
  syndicated programming, 30
Counter programming, 33, 168, 181, 209, 212
Crossley, A., 70, 73, 82, 84
Cross-sectional surveys, 89
Cultivation analysis, 55
Cumulative measures, 20, 35–37, 39, 61, 124, 134, 138, 147–150, 158, 181–182, 215–236, 240
  audience duplication, 171–173, 182, 223–228, 239
  inheritance effect, 38, 171–172, 226–227, 228, 234
  lead-in programming, 38, 135, 138, 171, 213, 214, 226–227
  mathematical modeling, 233–236
  dayparts and, 35–36, 147–150, 215–228
  meters and, 216

prime time and, 20
programming and, 228, 231–232

D

Databases, 81, 115, 119, 124, 137, 138–140, 149–150, 215, 241, 243
Dayparts, 153–157, 188–189, 194
  children's viewing, 55
  cumes and, 35–36, 147–150, 215–228
  defined, 5–7, 188
  early fringe, 30–31
  late night, 6–7, 112, 154
  radio, 28–29, 35, 132–134, 154, 155, 188–189, 216, 219
  see also Prime time
Demographic variables, 21, 32, 45–46, 49, 151–152, 162–163, 188, 196–198, 199–200, 239
  audience fragmentation, 169
  cable television, 240
  diaries and, 79, 113
  geography and, 22, 200; *see also* Zip codes
  historical use of, 73, 74
  income, personal, 70, 200
  location of television sets within home, 176–177
  Standard Metropolitan Statistical Areas, 125
  peoplemeters and, 80–81, 107–110, 240, 241
  programming, general, 27, 28–29, 32–33, 37, 61, 203–206
  radio listening and, 27, 28–29, 32–33, 37, 134, 157, 204–206
  sampling and, 97–98
  television viewing and, 33, 61, 121, 123, 151–152, 157, 209, 239, 240
  see also Age factors; Gender factors; Minority groups;
Designated market area (DMA), 12, 125
Diaries, 58–59, 74, 75, 77–79, 102–106, 114, 117, 239, 241
  Arbitron, 102, 103, 104
  cable television, 102, 106
  computer applications, 74, 112, 113
  demographics, 79, 113
  editing, 112–114
  electronic, 112
  error of measurement and, 58–59, 105–106, 112–113, 117–118

mechanical, 37, 241
meters and, 107, 108, 131, 138
radio, 29, 37, 74–75, 102, 104, 114, 132, 216
sampling, 78, 97, 117
telephone contacts, 78, 79, 104, 105
television, 55, 78–79, 102, 103, 104, 114

E

Early fringe daypart, 30–31
Economic and financial factors, 22–23, 43–51
　program choice models, 160–161, 162, 168–169
　see also Cost factors; Revenues; Supply/demand
Electronic Media Rating Council, 77, 141, 238, 239, 242
Error of measurement, 84, 88–89, 90–99, 102, 117–118, 130, 132, 146, 198–199, 207–208, 211–212
　diaries, 58–59, 105–106, 112–113, 117–118
　meters, 80–81, 107, 108–109
　remote control devices and, 106
　standard error, 92–94, 130, 211–212
　see also Sampling
ESPN, 195
Exposure, 20–21, 61, 62, 148, 168–174
　advertising, 220
　defined, 100–101, 223
　measurement of, 101–102, 146–165
　long-term, 183
　preference analysis, 56–58, 67, 153, 159–160, 162–163, 164, 165, 181, 182–183, 209
　selective, 55, 161
　see also Audience loyalty; Cumulative measures; Gross measures

F

Family viewing, 163
Federal Communications Commission (FCC), 52, 55–56, 58, 59–60
　cable television, 58
　children's television, 53
　licensing, 12
　ownership rules, 57–58
Federal Trade Commission, 52
Films and film industry, 30, 53, 60–61, 68, 160
FM radio, 10–11, 26, 166, 243
Focus groups, 28
Forecasts, *see* Projections
Format, *see* Programming

G

Gallup, G., 71, 75
Garrison, G., 77–78
Gender factors, 31, 73, 105, 113, 121, 132–134, 146, 162, 181, 188, 196–198, 199–200, 203, 229
Geographic variables, 11–12, 21–22, 45
　demography and, 22, 200; *see also* Zip codes
　Standard Metropolitan Statistical Areas, 125
　total survey area, 125, 131, 134
　See also Local markets
Glasser, G., 81–82
Gratificationist theory, 161–162, 174–175
Gross measures, 19, 20, 23, 147–150, 157, 170, 180, 185–214
　CPM/CPP, 23, 46, 48, 147, 186, 192–193, 199, 201–202
　households using television (HUT), 18, 19, 121, 130, 137, 153, 154, 157, 188, 190–191, 194, 208–209, 239
　person using television (PUT), 18, 130, 188, 190, 208–209, 214
　and cumulative measures, 228, 229

H

Harris Committee, 77
Historical perspectives
　A.C. Nielsen, 72, 73, 76–77, 78, 80, 81, 82, 107
　advertising, 5, 69–70
　Arbitron, 79, 80
　demographic data, 73, 74
　government policy, 43, 52–53, 56, 58, 59, 128
　gratificationist theory, 162
　meters, 73, 75–77, 79, 80, 106

SUBJECT INDEX

programming, 25, 39, 51, 70, 78
radio, 10, 26, 28, 39, 42–43, 67–74, 81
research methodology, 67, 69–81
social psychological effects of broadcasting, 42–43, 52, 54–55
television, 165
cable-broadcast competition, 59–60
programming, 25, 51
video cassette recorders, 175
Hooper, C., 71–73, 74, 78, 79, 82, 111
Households using television (HUT), 18, 19, 121, 130, 137, 153, 154, 157, 188, 190–191, 194, 208–209, 239

I

Independent stations, 26, 33, 48, 51, 53, 60, 157, 170
Independent Television Association, 53, 60
Indices, 199, 207–208, 228
average quarter hour (AQH), 124, 134, 190–192, 196, 199, 217, 218, 242
channel loyalty, 227
CPM/CPP, 23, 46, 48, 147, 186, 192–193, 199, 201–202, 239
households using television (HUT), 18, 19, 121, 130, 137, 153, 154, 157, 188, 190–191, 194, 208–209, 239
mail response as audience index, 68–69
persons using radio, 196
persons using television (PUT), 18, 130, 188, 190, 208–209, 214
time spent listening (TSL), radio, 34–35, 217–218, 228–229, 231
time spent viewing, television, 231–232
see also Cumulative measures; Gross measures
Individual factors, 174–177, 178, 180, 182, 213
vs mass behavior, 146–147, 178
see also Preference analysis
Infrared technology, 109
Inheritance effect, 38, 171–172, 226–227, 228, 234
lead-in programming, 38, 135, 138, 171, 213, 214
Interviews, *see* Personal interviews; Telephone methods

K

Karol, J., 71
Klein, P., 152–153, 209

L

Late night daypart, 6–7, 112, 154
Law
see Legal issues
Lazarsfeld, P., 43, 74
Lead-in programming, 38, 135, 138, 171, 213, 214, 226–227
inheritance effect, 38, 171–172, 226–227, 228, 234
Least objectionable program (LOP), 153, 164, 209
Legal issues
antitrust law, 58, 59
Cable Policy Act of 1984, 58
Constitutional law, 56, 58
copyright, 52, 58, 60–61
offensive language, 55
ratings practices, 77, 128
see also Litigation
Local markets, 4, 11–14, 119, 124–133, 151
area of dominant influence (ADI), 12, 124–125, 130, 131, 188, 200, 261–263
cable television, 14, 56, 58, 59–60, 114–115, 116, 151, 167, 241
designated market area (DMA), 12, 125
Metropolitan Statistical Areas, 125
news and public affairs, 56, 213–214
Nielsen Station Index (NSI), 76, 82, 124, 130, 138, 239–240
radio, 18, 26–29, 74, 76, 79, 81, 102, 120, 124, 131–133, 154, 166, 242–243
television, 18, 55–56, 57–58, 59–60, 76, 102, 124–131, 239, 241
see also Syndicated programming
Location of television sets within home, 176–177
Longitudinal studies, 89, 109, 134, 206
time series analysis, 212
see also Cumulative measures

M

Mail
diaries, 78, 105

response as audience index, 68–69
Marketing and Markets
  barter markets, 4, 167
  national markets, 4, 13
  opportunistic market, 8
  promotional campaigns, 68–69, 165
  scatter market, 7–8
  station representatives, 13–14, 119
  upfront market, 7, 210, 238
  see also Advertising and advertisers; Local markets
Market penetration, 16–19, 31–32, 49–50, 165–166, 170–171
  cable television, 8–10, 165–166, 169, 170, 181, 202
  radio, 27, 165–166
  share, defined/applied, 18–19, 157, 190–191, 201–202, 210, 213, 228
Market segmentation, see Behavioral factors; Demographic variables; Geographic variables
Mathematical models, 212, 232–236
Mechanical diaries, 37, 241
Memory, recall, 70, 72, 73, 74, 82, 110–112, 216
Meters, 82, 106–108, 115, 131, 216–217
  audimeters, 76, 79, 106
  cost factors, 82, 107, 140–141
  computer applications, 108, 109–110
  cumes, 216
  diaries and, 107, 108, 131, 187
  editing, 113
  error of measurement, 80–81, 107, 108–109
  history of use, 73, 75–77, 79, 80, 106
  peoplemeters, 80–81, 107–110, 115, 120, 131, 138, 173, 240, 241
  personal interviews, 73–75, 241
  sampling, 97, 108, 187
  telephone access to, 79, 107, 107–108
  television, 80–81, 107–110, 113, 115, 120, 131, 138
Methodology, 67–118
  Arbitron, 86–87, 102, 114
  historical perspectives, 67, 69–81
  longitudinal studies, 89, 109, 206
  Nielsen, 86, 120
  see also Cumulative measures; Diaries; Error of measurement; Gross measures; Meters; Telephone methods; Sampling; Statistical analysis

Metromail, 85–86
Metropolitan Statistical Areas, 125
Metzger, G., 81, 82
Minority groups, 74, 152, 231
  Asian Americans, 152
  Black Americans, 97, 105
  Hispanics, 8, 61, 73, 97, 152, 228–229
  sampling of, 98
Models
  mathematical models, 212, 232–236
  program choice models, 160–161, 162, 168–169, 177–183
Motion Pictures Association of America, 53, 60–61
Movies, see Films and film industry
Music, 26, 61, 68, 74, 75, 114

N

National Association of Broadcasters (NAB), 53, 60
National Cable Television Association (NCTA), 53, 60
National markets, 4, 13, 134
  see also Networks
NBC (National Broadcasting Company), 70, 73, 75, 78, 121, 152–153
Network Radio Audiences to All Commercials, 123–124
Network Radio Audiences to All Commercials Within Programs, 123–124
Networks
  affiliates, 30–31, 33, 48, 114, 140, 167, 170–171
  cable television, 8–10, 47, 52, 57, 59–60, 157, 165, 169, 188, 195, 228, 240
  cable television, competition with broadcast, 47, 52, 57, 59–60, 157, 165, 169
  as consumers of ratings, 119, 120–124
  cost of rating data, 140
  image of, 25
  loyalty to, 172
  radio, 10–11, 28, 123–124
  revenues, 44
  syndication compared to, 15–16
  television, 44, 107, 115, 120, 167, 120–124, 167–172, 239
    advertising, 5–10, 11, 20, 44
    programming, 25, 29
    upfront market, 7, 210, 238

SUBJECT INDEX

see also Affiliates, network; specific networks
News programming, 6, 33, 35, 56, 75, 157, 181, 194, 204–205, 213–214, 232
Newspapers, see Print media
Nielsen, A. C., Jr., 56–57
Nielsen, A. C., Sr., 76, 85
Nielsen Homevideo Index, 240
Nielsen Media Research, 37, 216–217
Nielsen Radio Index (NRI), 76, 82, 134
Nielsen Station Index (NSI), 76, 82, 124, 130, 138, 239–240
Nielsen Syndication Service (NSS), 135–137, 240
Nielsen Television Index (NTI), 76, 80, 116, 120–123, 137, 189–190, 221–223, 238–239
NSI Plus, 138

O

Off-network programming, 29–30
Opportunistic market, 8

P

Panel studies, 89
The People Look at Television, 151
Peoplemeters, 80–81, 107–110, 115, 120, 131, 138, 173, 240, 241
Personal interviews, 73–75, 241
Persons using radio (PUR), 196
Persons using television (PUT), 18, 130, 188, 190, 208–209, 214
Pocketpiece, see Nielsen Television Index
Policy issues, 43, 44, 51–62
  see also Federal Communications Commission; Government role
"Postbuy Reporter," 130
Preference analysis, 56–58, 67, 153, 159–160, 162–163, 164, 165, 181, 182–183, 209
  least objectionable program, 153, 164, 209
Prediction, see Projections
Prime time, 20, 29, 34, 40, 47, 157, 181, 239
  cumulative ratings, 20
  day-of-the-week factors, 153–154
  defined, 5–6
  Nielsen Television Index (NTI), 121

Princeton Radio Research Project, 43
Print media, 42, 46, 70, 83
Probability, 85
Program preferences, see Preference analysis
Programming, 3, 25–41, 159–160, 164–165, 166–169
  affiliates, network, 33, 167
  audience loyalty and, 172–173, 230, 232
  cancellation, 39–40
  children's, 55, 163
  cost factors, 39–41, 169
  counter programming, 33, 168, 181, 209, 212
  cumulative measures, use of, 228, 231–232
  demographics of, 27, 28–29, 32–33, 37, 61, 203–206
  diaries, 78
  history, 25, 39, 51, 70, 78
  inheritance effect, 38, 171–172, 226–227, 228, 234
  lead-in, 38, 135, 138, 171, 213, 214, 226–227
  radio, 26–29, 32–33, 34–37, 114, 204–206, 228–229
  Spanish-language, 51, 61, 73, 97, 152, 228–229
  television, 29–31, 33, 35, 37, 38, 39–41, 51, 81, 114–115, 167–169, 180–183, 231–232
  see also Dayparts; Syndicated programming; *specific program types*
Projections, 115, 117–118, 134
  audiences, 44, 50, 130, 186–187, 207
  and explanation, 208–214, 232–236
Promotional campaigns, 68–69, 165
Psychology, see Individual factors; Preference analysis; Social psychology
Public broadcasting, 33–34, 101, 206, 231
Public opinion, 53, 83–84
The Pulse of New York, 74–75, 77, 79, 81

Q

Qualitative ratings, 101, 165–166
Quarter hour maintenance, 34 *also see Time spent listening*

R

RADAR (Radio's All Dimension Audience Report), 81–82, 110–111, 113, 123, 124, 154, 242–243
RADAR On-Line, 124, 243
Radio, 10, 76, 81–82, 133–134, 148–149, 216–219, 241
  advertising, 4–5, 10–11, 17–18, 47–48, 70, 71, 81
  affiliates, network, 114
  AM/FM, 10–11, 27, 132, 134, 166, 205, 243
  Arbitron, 28, 29, 81, 102, 124, 131–133, 216, 240, 241
  audience loyalty, 173, 182–183, 230
  audience size determination, 17–18, 35–37, 69–73, 190
  Birch Radio, 80, 81, 110, 131–132, 134, 242
  in cars, 112
  commercial time, 47–48
  computer applications, 28, 37, 124, 134, 239, 240, 241, 242
  daily listening, 157
  day-of-the-week factors, 154
  dayparts, 28–29, 35, 132–134, 154, 155, 188–189, 216, 219
  demographics, 27, 28–29, 32–33, 37, 134, 157, 204–206
  diaries, 29, 37, 74–75, 102, 104, 114, 132, 216
  historical perspectives, 10, 26, 28, 39, 42–43, 67–74, 81
  local, 18, 26–29, 74, 76, 79, 81, 102, 120, 124, 131–133, 154, 166, 242–243
  networks, 10–11, 28, 123–124
  Nielsen Radio Index (NRI), 76, 82
  persons using radio (PUR), 196
  programming, 26–29, 32–33, 34–37, 114, 204–206, 228–229
  RADAR, 81–82, 110–111, 113, 123–124, 154, 242–243
  telephone methods, 112
  time spent listening (TSL), 34–34, 217–218, 228–229, 231
"Radio Spot Buyer", 134
Radio Usage, 123
RCA, 75
Recall, 70, 72, 73, 74, 82, 110–112, 216
Recycling, radio, 35
Remote control devices, 106, 177

Repeat viewing, 173–174, 227, 235–236, 239
Report on Syndicated Programs, 135, 136, 137
Research, 67–83
  academic, 43, 53–55, 62, 71, 73, 74, 75–76, 77–78, 164
  advertising, 16–23
  economic, 45–51
  history, 67, 69–79
  policy, 54–62
  products of, 119–141
  programming, 31–41
  social psychological effects of broadcasting, 42–43, 54–62, 67
  see also Methodology
Revenues, 47, 43–44
  advertising, 4–5, 47
  programming and, 39–41
  ratings and, 47–50
Ridgeway, R. R., 79
Robinson, C., 75
Roslow, S., 74
Roster recall, 74

S

Sampling, 85–99, 148, 198–199, 207, 211–212
  diaries, 78, 97, 117
  meters, 97, 108, 187
  minorities, 98
  radio, 96
  sweeps, 79, 82, 102, 127–130, 131–132, 211
  telephone, 86–87, 97, 123
ScanAmerica, 110
Scatter market, 7–8
Seasonal patterns, 153
  advertising time transactions, 7–8
  media use, 153
  see also Sweeps
Seiler, J., 78–79
Selective exposure, 55, 161
Sex differences, *see* Gender factors
Silverman, F., 29
Social psychology, 22, 42–62, 67
  attitudes, 4, 56, 106, 163
  audience loyalty, 19–21, 34, 61–62, 172–174, 227, 230, 232, 234

SUBJECT INDEX

gratificationist theory, 161–162, 174–175
group viewing, 163
historical perspectives, 42–43, 52, 54–55
individual *vs* mass behavior, 146–147, 178
personalities, celebrities, 26, 27–28, 194, 207
preference analysis, 56–58, 67, 153, 159–160, 162–163, 164, 165, 181, 182–183, 209
program choice models, 160–161, 162, 168–169, 177–183
public opinion, 53, 83–84
selective exposure, 55, 161
television, effect of, 53, 54–55
television set proliferation and, 176–177
Spanish-language programming, 51, 61, 73, 97, 152, 228–229
Spot advertising, 47, 130
 local, 13
 national, 4, 11, 13
"Spotbuyer," 130
Standard error, 92–94, 130, 211–212
Standard Metropolitan Statistical Areas *see* Metropolitan Statistical Areas
Standards
 accreditation, Electronic Media Rating Council, 141, 238, 239, 242
 diaries, 113
 methodological, 101–102, 112, 113
Stanton, F., 42, 43, 67, 75
Starch, D., 71, 73
Station representatives, 13–14, 119
Statistical Research, Inc., 80, 81, 82, 123–124, 242–243
Statistical analysis, 85, 99–112
 aggregation, 11
 cross-tabulation, 224–226
 editing, 112–114
 of mass behavior, 178
 mathematical models, 212, 232–236
 peoplemeters, data bias, 80–81
 time series analysis, 212
 see also Cumulative measures; Error of measurement; Gross measures; Indices; Sampling
Steiner, G., 151, 151
Stengel, C., 29
Supply/demand
 audience value and, 45–46
 off-network programming, 30

Sweeps, 79, 82, 102, 127–130, 131–132, 211
Syndicated Program Analysis, 135
Syndicated programming, 14–16, 25, 37–38, 135–137, 140
 advertisers and advertising, 11, 14–16, 37–38
 barter, 14–16, 37–38, 167
 as consumers of ratings, 119
 exclusive rights, 60–61
 government policy, 60
 radio, 11, 28
 ratings products, 135–137
 revenues, 44
 television, 26, 29–30, 39, 135–137, 240

T

Target audience, 32, 199, 200
Tartikoff, B., 29
Teenagers, *see* Adolescents
Telephone methods, 59–73
 diaries and, 78, 79, 104, 105
 meters, access to, 79, 107, 107–108
 random digit dialing, 87, 123
 recall, 70, 72, 73, 82, 110–112, 216
 sampling, 86–87, 97, 123
Tele-Que, 79
Television, 148–149, 240–241
 advertising revenues, 4–5
 Arbitron, 82, 124–131, 135, 240–241
 audience size determination, 17–18, 152–154, 157, 180, 186–188, 190–191, 207
 cable television, competition, 47, 52, 57, 59–60, 157, 165, 169
 computer applications, 37, 130, 131, 239, 240, 241
 daily viewing, 157–158
 day-of-the-week factors, 153–154
 demographics of, 33, 61, 121, 123, 151–152, 157, 209, 239, 240
 diaries, 55, 78–79, 102, 103, 104, 114
 independent stations, 26, 33, 48, 51, 53, 60, 157, 170
 local, 18, 55–56, 57–58, 59–60, 76, 102, 124–131, 239, 241
 meters, 80–81, 107–110, 113, 115, 120, 131, 138
 network, 44, 107, 115, 120, 167, 120–124, 167–172, 239

Nielsen Station Index (NSI), 76, 82, 124, 130, 138, 239–240
Nielsen Television Index (NTI), 76, 80, 115, 120–123, 137, 189–190, 221–223, 238–239
  out-of-home viewing, 106
  programming, 29–31, 33, 35, 37, 38, 39–41, 51, 81, 114–115, 167–169, 180–183, 231–232
  repeat viewing, 173–174, 227, 235–236, 239
  revenues and audiences, 47–50
  social psychological effects, 53, 54–55
  sweeps, 79, 82, 102, 127–130, 131–132, 211
  syndicated programming, 26, 29–30, 39, 135–137, 240
  telephone methods, 112
  time spent viewing, television, 231–232
  UHF *vs* VHF, 167, 180
  see also Cable television; Households using television; Persons using television; Video cassette recorders
Television Audience, 123
Television Audience Assessment, 101
Theories
  defined, 145–146
  exposure, 100
  gratificationist, 161–162, 174–175
  least objectionable program, 153, 164, 209
  Marxist economics, 44, 46
  measurement, 112
  neoclassical economics, 44, 45–46
  program choice models, 160–161, 162, 168–169, 177–183
  program scheduling, 38, 114–115, 168, 180–181
  sampling, 85–99 (passim)
Time series analysis, 212
Time spent listening (TSL), radio, 34–35, 217–218, 228–229, 231

Time spent viewing, television, 231–232
Total survey area (TSA), 125, 131, 134
Trend studies, 89, 134
"TV Maximizer," 130
Turnover, 34–35

U

Universities and colleges, *see* Research academic
Upfront market, 7, 210, 238
Urban areas, *see* Metropolitan Statistical Areas

V

VCR Tracking Report, 123
Video cassette recorders, 48, 94, 113–114, 175–176, 240
  diaries, 106
  Nielsen Homevideo Index, 240
Viewer Tracking Analysis, 37
Violence, 52, 55, 61

W

White, P., 71
Woodruff, L.F., 75, 76

Y

Young and Rubicam, 71

Z

Zip codes, 22, 37, 139, 200, 202